DATE DUE

BRODART   Cat. No. 23-221

OXFORD GEOGRAPHICAL AND ENVIRONMENTAL STUDIES

General Editors: Gordon Clark, Andrew Goudie, and Ceri Peach

Culture and the City in East Asia

Edited by

Won Bae Kim, Mike Douglass, Sang-Chuel Choe, and Kong Chong Ho

CLARENDON PRESS · OXFORD
1997

Oxford University Press, Great Clarendon Street, Oxford OX2 6DP

Oxford New York

Athens Auckland Bangkok Bogota Bombay
Buenos Aires Calcutta Cape Town Dar es Salaam
Delhi Florence Hong Kong Istanbul Karachi
Kuala Lumpur Madras Madrid Melbourne
Mexico City Nairobi Paris Singapore
Taipei Tokyo Toronto Warsaw

and associated companies in
Berlin Ibadan

Oxford is a trade mark of Oxford University Press

Published in the United States by
Oxford University Press Inc., New York

British Library Cataloguing in Publication Data
Data available
ISBN 0-19-823358-2

Library of Congress Cataloging in Publication Data
Culture and the city in East Asia / edited by Won Bae Kim . . . [et al.].
p. cm.—(Oxford geographical and environmental studies)
Includes bibliographical references and index.
1. Urban geography—East Asia. 2. Human geography—East Asia.
3. Cities and towns—East Asia—Growth. 4. City planning—East
Asia. 5. East Asia—Social policy. 6. East Asia—Economic policy.
I. Kim, Won Bae. II. Series.
GF651.C85 1997 96-24109
307.76′095—dc21
ISBN 0-19-823358-2

1 3 5 7 9 10 8 6 4 2

Typeset by Best-set Typesetter Ltd., Hong Kong
Printed in Great Britain by
Bookcraft (Bath) Ltd., Midsomer Norton, Somerset

EDITORS' PREFACE

GEOGRAPHY and environmental studies are two closely related and burgeoning fields of academic inquiry. Both have grown rapidly over the past
two decades. At once catholic in its approach and yet strongly committed to
a comprehensive understanding of the world, geography has focused upon
the interaction between global and local phenomena. Environmental studies, on the other hand, have shared with the discipline of geography, an
engagement with different disciplines addressing wide-ranging environmental issues in the scientific community and the policy community of great
significance. Ranging from the analysis of climate change and physical
processes to the cultural dislocations of postmodernism and human geography these two fields of inquiry have been in the forefront of attempts to
comprehend transformations taking place in the world, manifesting themselves in a variety of separate but interrelated spatial processes.

The new 'Oxford Geographical and Environmental Studies' series aims
to reflect this diversity and engagement. It aims to publish the best and
original research studies in the two related fields and in doing so, to demonstrate the significance of geographical and environmental perspectives for
understanding the contemporary world. As a consequence, its scope will be
international and will range widely in terms of its topics, approaches, and
methodologies. Its authors will be welcomed from all corners of the globe.
We hope the series will assist in redefining the frontiers of knowledge and
build bridges within the fields of geography and environmental studies. We
hope also that it will cement links with topics and approaches that have
originated outside the strict confines of these disciplines. Resulting studies
will contribute to frontiers of research and knowledge as well as representing individually the fruits of particular and diverse specialist expertise in the
traditions of scholarly publication.

Gordon Clark
Andrew Goudie
Ceri Peach

PREFACE

THE East–West Center in Honolulu and the Seoul Development Institute initiated a collaborative research project on culture and built environment in East Asian cities in October 1993. A major workshop, held in Honolulu in August 1994, discussed the preliminary results of this comparative research. All the participants agreed on the need to explore the overriding importance of culture in urban transformations.

Despite recent interest in culture and institutions in development economics and urban studies, there is little research on the relationship between culture and the urban process in Asia. To help fill this gap, the project emphasized case studies in which country collaborators would look into the interplay of culture, social processes, and the built environment in specific historical and cultural contexts.

The Introduction by the editors and Part I of this book cover the central themes of this project and link them to recent developments in urbanization and urban morphology in East Asia. Chapter 2 sets the urban transformation in a cultural and historical context. Chapter 3 is devoted to the theory of urban process, incorporating the realities of East Asia. The fourth chapter examines the urban transition in the cities of East Asia in terms of work, space, and place.

Part II on Culture and the Built Environment includes a brief introductory chapter and six subsequent chapters that investigate the interplay of culture, socio-economic and political processes, and the built environment in specific contexts and at specific urban sites. The introductory chapter provides a background of the methodology for the case studies and a general overview of the selected cities. The following six chapters are stories and explanations of the formation and transformation of urban landscapes in six capital cities of East Asia: Beijing, Seoul, Tokyo, Hanoi, Hong Kong, and Singapore. They provide the basis contrasting localized experiences and making general observations about urban transformations in East Asia. The final chapter reflects on the likely future of East Asian cities, noting the increasing significance of culture in their long-term growth. Ways to strengthen the social and cultural creativity of cities are also considered.

The collaborative project on which this book is based was designed to expand available information on the past and present experiences of cities of East Asia and to provide a perspective for understanding the role of culture in urban transformations. An interest in articulating the interplay between culture, social process, and urban form in the cities of East Asia has also drawn attention to comparing the East Asian experience with that of other continents and regions. While there is much that is specific to each of the cities described here, there are also some commonalities among the cities in East Asia that distinguish them from Western cities. (Western cities are not monolithic either but this contrast serves the purpose of comparison at the high level of aggregation.) This project began with the working hypothesis that all of the cities it covers under the rubric of 'East Asia' share a common tradition of Confucianism, although there are variations in the way that Confucian tradition has been played out in different local contexts. Other common dimensions include the fact that most East Asian cities were latecomers to capitalist development, that they have been considerably influenced by the West, and that they all experienced direct or indirect colonial influence or were, in fact, colonial creations.

Furthermore, unlike the situation in the West, contemporary East Asian societies have all experienced a compressed process of urbanization and economic development with pervasive state intervention in economic and social affairs. Apart from the two socialist nations of China and Vietnam, East Asian countries have gone through a rapid transformation from an agrarian to a wage-based industrial economy, accompanied by a massive transfer of population from the countryside to the city. Indeed, the pace and scale of urbanization in Japan and Korea is unprecedented in world history. Cities everywhere in East Asia have been expanding rapidly, playing a central role in national economic growth, and producing a similar set of urban problems. These cities have also been rapidly incorporated into global trade and communications networks as East Asian countries have become integrated in new ways into the global capitalist system. Contemporary urbanization and urban processes in East Asia are thus the result of national social transformations in an expanding global capitalist system.

While sharing these common characteristics, each society in East Asia has taken its own specific route. Developmentalism prompted by the state did not produce uniform outcomes across all societies. Even the two city-states of Hong Kong and Singapore, both colonial creations, did not follow the same developmental model. History and geography played out differently in the development of socio-cultural, economic, and political structures. A focus on these variations highlights the strength and resilience of local socio-cultural institutions as they filter external influences, manage economic and other crises, and create varied modes and paths of development.

This book intends to bring out both the differences among East Asian cities and the commonalties that they share. Case studies were undertaken to examine the interplay between the city and social change. In this interplay, we take the position that culture matters greatly in societal processes, including urbanization. Departing from the strictly economic treatment of urbanization that has tended to dominate urban geography, we consider culture to be a key dimension of social change as well as urban transformation in all societies.

To explore the role of culture in urban transformation, specifically in the production and reproduction of the built environment, the case studies in this book have been based on specific sites, issues, or events. These specific analyses are intended to illustrate part of the complex process involved in the relationships between social continuity and change and the experiences of power that both create and are contingent upon the functioning and symbolic meanings of the urban ensemble. Each case study is designed to tell a story, incorporating a description of the general dynamics that have shaped the urban form of each particular city and providing a basis for comparison with other cities.

As reflected in the diverse backgrounds of the contributors to this volume, urban transformations in East Asia are not interpreted from a single perspective. Although Western research methodology has been employed, a unified view has not been adopted, either Western or Asian. Instead, each chapter has its own structure and logic, emphasizing different aspects of urban processes. The chapters on specific cities contain the stories and interpretations of urban transformation by local scholars. We hope that this book will contribute to the knowledge and understanding of policy-makers and academics concerned with the future of cities in East Asia.

We would like to acknowledge the Program on Population of the East–West Center and the Seoul Development Institute for their financial and organizational resources, which made this project possible. We would also like to thank researchers from the Seoul Development Institute and the East–West Center and academics from the University of Hawaii for their keen interest in the project. In particular, we would like to acknowledge the support of Dr Sae-Wook Chung (president of the Seoul Development Institute), Dr Andrew Mason (director of the East–West Center's Program on Population), Dr Wimal Dissanayake, and Dr Yok-shiu Lee of the East–West Center, and Dr Wang Feng, Dr Sen-dou Chang, Dr Hagen Koo, Dr Truong Buu Lam, Dr Harry Lamley, Dr Mary McDonald, Dr Alvin So, and Dr Stephen Yeh from the University of Hawaii for their useful comments on the papers presented at the workshop. Finally, we would like to thank Sidney Westley, Mary Ann Lee, and Connie Kawamoto for their valuable assistance in compiling and editing the manuscript.

M. D.
W. B. K.

CONTENTS

LIST OF FIGURES

LIST OF TABLES

LIST OF CONTRIBUTORS

SANG-CHUEL CHOE is a professor of urban and regional planning at the Graduate School of Environmental Studies, Seoul National University. He has served as dean of GSES, SNU, and president of the Seoul Development Institute. He has contributed many papers in a series of books published by the United Nations Centre for Regional Development, the United Nations University, and the East–West Center on the topics of urban and regional development, urban planning history, and land-use planning. Dr Choe is currently working on a book titled 'Seoul: From the Capital of Hermit Kingdom to Modern Metropolis'.

MIKE DOUGLASS is professor and chair of the Department of Urban and Regional Planning at the University of Hawaii and is a senior fellow at the East–West Center. He holds a Ph.D. in urban planning from the University of California at Los Angeles (UCLA). He has served as a visiting scholar at Tokyo University, research staff member at the United Nations Centre for Regional Development in Nagoya, a Peace Corps volunteer in Korea, and a consultant on national, urban, and regional policy formulation in Indonesia and Thailand. He has taught at Doshisha University in Kyoto, Tong-A University in Pusan, and Thammasat University in Bangkok as well as at the Institute of Social Studies in The Hague, the University of East Anglia in Norwich, and at UCLA.

KONG CHONG HO is a senior lecturer in the Department of Sociology at the National University of Singapore. He holds a Ph.D. in sociology from the University of Chicago. He is an associate editor of the *Southeast Asian Journal of Social Science* and the Sociology Working Papers Series. His research interests include urban studies, information technologies, development, and change. Dr Ho co-edited *Explorations in Asian Sociology* and is currently working on a book comparing the development experiences of Hong Kong and Singapore.

WON BAE KIM is a senior fellow at the Korea Research Institute for Human Settlements and has an affiliation with the Program on Population of the East–West Center. He holds a Ph.D. from the University of Wisconsin. Dr Kim is interested in migration, urbanization, and regional development in Asia. He is co-editor of a book, *Asian NIEs and the Global Economy: Industrial Restructuring and Corporate Strategy in the 1990s* (Johns Hopkins University Press), and has published journal articles and book chapters on urban, regional, and development studies.

REGINALD YIN-WANG KWOK is a professor of Asian studies and urban and regional planning at the University of Hawaii at Manoa. He holds a Ph.D. in urban and regional planning from Columbia University. He sits on the Advisory Board of the International Research Foundation for Development and the Editorial Board of Built Environment. His research interests include industrialization and urbanization in China and East Asia, political economy of international and national development, development and regional economics, spatial planning, and urban design. He co-edited *Chinese Urban Reform: What Model Now?* and *Regional Development in Asia*, co-wrote *The Sek Kip Mei Syndrome: Economic Development and Public Housing in Hong Kong and Singapore*, and wrote *General Theories of Urban Planning* (in Chinese). Currently, he is working on a book entitled *Hong Kong–Guangdong Link: Partnership in Flux*.

SI-MING LI is a professor of geography at the Hong Kong Baptist University and is currently visiting professor of geography at the National Taiwan University. His main research and teaching interests focus on urban housing markets, urban transport, and China's population mobility and regional development. His publications include a book on Hong Kong's urban issues (co-authored with Fu-lai Yu), a volume on China's social development (co-edited with Yat-Ming Siu and Tai-Kei Mok), and numerous articles published in academic journals.

TRINH DUY LUAN is a senior researcher in the Institute of Sociology at the National Centre for Social Sciences (NCSS) of Vietnam. He is a Ph.D. candidate in sociology at the Russian Academy of Sciences. He is currently the head of the Department for Urban and Community Studies at NCSS. His research interests include urban studies, social change in contemporary urban Vietnam, and urban housing. He also teaches urban sociology at Hanoi University.

TERRENCE G. MCGEE is a professor at the Department of Geography and the Institute of Asian Research, University of British Columbia. He holds a Ph.D. in geography from Victoria University of Wellington, New Zealand. He has carried out research on urbanization in Asia for almost forty years. His books include *The Southeast Asian City* (1967), *Hawkers in Hong Kong* (1976), and *Theatres of Accumulation* (1985) with Warwick Armstrong and he co-edited *The Extended Metropolis in Asia* (1991).

TAKASHI MACHIMURA is an associate professor of sociology at Hitotsubashi University in Tokyo. He recently visited UCLA as a Fulbright scholar to conduct research on the cultural basis of global cities. He is the author of *The Structural Change of a Global City: Urban Restructuring of Tokyo* (University of Tokyo Press) and several publication on urban issues.

HYUNGMIN PAI is an assistant professor in the Department of Architecture at Seoul City University where he teaches modern architecture and design. He holds a Ph.D. in architecture, art, and environmental studies from the Massachusetts Institute of Technology. His main interests are in the history, theory, and criticism of architecture and urbanism. He is at present working on a book titled 'From the Portfolio to the Diagram, Architecture, Discourse, and Modernity in America', to be published by MIT Press in 1996.

GISÈLE YASMEEN has completed her Ph.D. in the Department of Geography at the University of British Columbia. Her dissertation is entitled 'Gendered Access to Public Space: Cooked-Food Sellers in Bangkok'. She holds an MA in geography from McGill University where her thesis research focused on housing co-operatives in Montreal.

ZIXUAN ZHU is a professor of city planning and urban design at the School of Architecture, Tsinghua University, Beijing. Being a consultant of city planning and historical preservation to the Beijing municipal government, he is a chief designer of the conservation and rehabilitation of the Shi-sha-hai Historic Area. Professor Zhu's main interests focus on urban design and historic preservation and rehabilitation.

1

Introduction

Won Bae Kim, Mike Douglass, and Sang-Chuel Choe

The focus on culture in this book represents a departure from much of the contemporary study of urbanization. Previous work on cities in East Asia has all too often assumed a generic 'Third World' or 'Asian' model of development that fails to consider substantial differences in historical experience. Culture is here understood as expected ways of behaviour that are encoded in social, political, and economic institutions. It provides a normative context within which social behaviour is instituted and acted out (Agnew *et al.* 1984).

Culture is dynamic, in a constant state of flux (Rotenberg 1993). It is integrated into all dimensions of human experience, including the spatial dimension. Yet because culture is intertwined with relations of power, its deeper aspects are often obscured by its more visible social, economic, and political manifestations. This book emphasizes the cultural dimension of urban development by revealing the central importance of cultural factors in social transformation as they are played out in urban space. It counters the view that culture is only relevant to the past or should be equated with 'tradition'. By investigating a part of the world with a long urban history, this book demonstrates both cultural change and continuity. Equally important, it shows how the pervasive influence of culture in the social, political, and economic sphere can help shape the future of Asia and its cities.

Although the importance of cultural differences is still debated, a consensus is emerging that the nations of the world may follow diverse development paths, rather than a single linear progression toward Western capitalism (Berger and Hsiao 1988; Murakami 1992). At a very general level, a set of common features can be identified that distinguishes the countries of East Asia from the Western model. East Asia is defined here to include the countries and societies which share Confucianist ideals and tradition, such as China, Japan, Korea, Hong Kong, Singapore, Taiwan, and Vietnam. Communitarianism, the emphasis on social relations rather than individualism, is one common feature that is often linked to the continuity of Confucianist ideals and institutions in East Asia (Pye 1985; Vogel 1991).

More recently, state-led or state-assisted developmentalism has been noted in East and South-East Asia as a critical feature of rapid economic growth (Amsden 1989; Haggard 1990; Wade 1990).

East Asians are thought to place a high value on loyalty and trust in social relations. Social and economic status have been differentiated according to an ideologically determined hierarchy, usually in the descending order of the literati—the farmer—the craftsman—the merchant. In contrast to capitalist development in the West, power and wealth tend to have been not so closely associated until very recently (Pye 1985; Sakakibara 1993). In many ways the rich in East Asia may enjoy less respect and social status than their counterparts in the West, although the advancement of capitalism has been gradually changing this perception in major cities of East Asia.

While the merits of these features may be debated, they add a useful dimension to studies of development, not only differentiating East from West but also distinguishing specific societies within East Asia. Each of the case studies in this book demonstrates the uniqueness of cultural change and its iterative feedback through social transformations in individual societies. Korean culture and history, for example, differs markedly from that of Japan or China. These differences are manifested in such areas as the role of the state in the economy, the maturity of civil society, and the built environment of cities. By extension, Korea's future will spring from a different social, economic, and political base from the development of its neighbours.

The list of contrasts between specific East Asian countries could be elaborated in more detail. The point is that cultural differences are too important to be treated as residual factors. Their importance increases as countries in the global community interact more frequently and become more interdependent.

This book does not focus on culture exclusively, but rather on the interface between culture and the city. Little research has been done on this topic in Asia, particularly with regard to placing the contemporary period of rapid economic growth and accelerated urbanization into the longer sweep of history. Published work has tended to focus on pre-colonial and colonial cities rather than on cities of the late twentieth century and has been dominated by the Western modernist perspective that overly generalizes the Asian experience (Basu 1985; Beg 1986; Costa *et al.* 1989; McGee 1967; Ross and Telkamp 1985; Smith and Nemeth 1986).

Since urban transformation occurs in specific locations, any analysis must take account of local history and culture. Unfortunately, urban studies has tended to neglect the specifics of local history in favour of a more generalized demographic and economic approach (Hauser *et al.* 1985; Fuchs *et al.* 1987). Research on specific cities has tended to focus on particular urban functions, with a fragmented perspective that is essentially economic (Richardson 1990; Rondinelli 1991).

Most contemporary analyses of urban processes are incomplete and largely ahistorical. This gap led the authors of this book to concentrate on the interface between culture and the city. The emphasis is not on culture in a unidimensional sense, but rather on the complex, multidimensional transformation of cities by social, political, and economic forces. Culture in this sense is a set of values, attitudes, and institutions that shapes social action just as action in turn shapes cultural change.

The definition of 'city' has proven to be just as elusive as that of 'culture'. Researchers have paid most attention to distinguishing rural and urban places in demographic and economic terms, namely according to population size, density, and non-agricultural functions (Goldstein and Sly 1975; Jones 1983). Underlying such definitions is an often implicit understanding that the emergence of cities and the expansion of urban systems is the spatial expression of fundamental transformations in scale, political configuration, levels of material welfare, and a wide array of social variables. It is important to distinguish between urbanization as part of a process of social change and the city as a site or geographic space in which these changes are acted out. While much of Western sociology, economics, and political economy—including modernization and capitalist transformation theories—have considered the real and imagined outcomes of urbanization, the city has only recently become a site for exploring the complex of forces at play. One reason why such studies have been rare is that the major academic disciplines have dismissed the importance of space in society; another reason has been the rejection of the city as a distinct category of analysis. This book joins other works that are reinserting considerations of geographical space into social analysis (Lefebvre 1991; Soja 1989) and recognizing the city as a focal point of social forces and, as such, worthy of analysis (Castells 1977, 1983).

As the spatial correlate of culture, the city has been associated with major transformations in world history. Western urbanization, for example, has been seen as an intrinsic feature of the evolution from imperial to democratic systems of governance, while cities in East Asia appear to reflect the oscillations between centralized and decentralized structures of government. Cities are likewise associated with the rise of markets and the transition from feudalism to capitalism.

Like social transformations, changes in the built environment of cities are rarely free of the past. For cultural, political, and economic reasons, a city's architecture, form, and physical development generally reflect some degree of inertia. Cities such as Beijing, Hanoi, Seoul, and Tokyo can be seen as an overlay of the present on an accumulation from the past that is also moving into the future. Under the contemporary surface, the past lives on, influencing the shape of the present and of the future. The city is also a node in a wider system of interaction where the indigenous meets the foreign and the global is articulated and interpreted at the local level. It is a home to numerous people, who often compete for power and space, creating distinctive spatial orders. Such spatial orders act in turn as a physical and symbolic parameter for social action (Lawrence and Low 1990).

Within this framework of the city as a place where the past meets the present, the indigenous meets the foreign, and different groups compete for power and space, this book examines the interplay between socio-cultural institutions and urbanization in East Asia. Six case studies of East Asian cities illustrate urban processes that are more open-ended and more historically and geographically specific than generally allowed by modernist or Marxist theory. Beijing, Hanoi, Seoul, Tokyo, Hong Kong, and Singapore are all located in one of the most dynamic regions of the world. They have all experienced rapid growth and are currently undergoing significant social, economic, and political transformation.

These cities play an increasingly important role in the Asia-Pacific region. As capital cities or city-states, they function as gateways to their countries and thus central points for tension between global and local interests and traditional and modern forces. The tension may appear to be sharper in what have been termed the 'orthogenetic' (in the sense of carrying an old culture forward) cities of Beijing, Seoul, Tokyo, and Hanoi than in the 'heterogenetic' (in the sense of creating an original mode out of more than one old culture) cities of Hong Kong and Singapore (Redfield and Singer 1954). In orthogenetic cities, the juxtaposition of old neighbourhoods with new commercial and industrial areas is often more striking than in the physically varied heterogenetic cities. For example, Hanoi—a city dominated by colonial architecture but now undergoing intensive redevelopment pressure—reflects the contrast between Vietnam's imperial and socialist regimes and the more recent introduction of a market economy.

Despite their social, political, and economic differences, all these cities tend to share a similar contemporary set of urban problems arising from rapid growth. Such problems include land speculation, an unstable real estate market, perennial shortages of public goods and services, sharp disparities between rich and poor in access to urban space, increasing conflict over urban space, and a deteriorating urban environment. In addition, rapid growth in the past decades has brought on an identity crisis, especially among higher-income cities. Just as industrialization in Japan, the Asian newly industrialized economies (NIEs), and now China, Vietnam, and South-East Asian countries has substantially departed from the stylized European and North American model, so the cities in these regions are developing along paths distinct from those of the West (Ginsburg *et al.* 1991).

The studies in this book look at cities as sites for social action and also as nodes in the global flow of resources, including commodities, capital, and information. The emphasis is on the built environment and its transformation. The interplay of culture, economics, and politics is discussed under three major themes: (1) culture and urban transformation; (2) function, symbolism, and power; and (3) the city in a new global order. This chapter will introduce these themes and sketch out an approach to analysing the relationship between culture and the city.

Culture and Urban Transformation

Culture, as a set of values and institutions, underpins the division of labour and the organization of power. Culture, as a set of beliefs and rules of behaviour, gives legitimacy to the organization of a society; and culture is manifested in the physical development of cities through space and time. Throughout history, ideologies and institutions have influenced the location of cities, their physical layout, the allotment of space to various segments of society, and even the construction and positioning of houses and other buildings (Wheatley 1972). Although these visible manifestations of culture have been increasingly influenced by market forces, other aspects of culture continue to play an important role in the transformation of urban space. Conversely, the built environment serves symbolic and ideological functions that underpin the cultural basis of status, power, and authority (Lawrence and Low 1990).

One basic assumption of this perspective is that the structuring of urban space is directed by evolving institutions that are, in turn, the embodiment of cultural norms and practices. These institutions also shape the division of labour, the regulation of power, and the definition of such basic elements of social exchange as trust and meaning (Eisenstadt and Shachar 1987).

The expression of urban form and structure in concrete political, geographical, and sociological settings is also influenced by international political and economic systems. While different actors, forces, and processes shape the basic premises of a society and its institutional framework, it is the political élite, such as the nobility and intelligentsia in pre-modern China, Korea, and Vietnam, that articulates the norms of the social, political, and economic order. It is the élite that also mobilizes, regulates, and controls society's resources. Control is not only exercised through political-military power and economic dominance but also through rules of social hierarchy and authority (A. Gramsci, quoted in Eisenstadt and Shachar 1987). Authority over the large territorial expanse required to sustain a city can rarely be maintained through violence or coercion alone. It is more often derived from popular acceptance of culturally defined hierarchies and, by extension, the central role of the capital city itself. The threat of force may underlie such popular assent, but this does not contradict the role of culture in the extension of territorial control from key urban nodes: rather, it demonstrates the intertwining of culture with other social and spatial dimensions.

The history of cities in East Asia is closely related to the articulation of cultural norms that define the social order. Yet the institutionalization of any social order inevitably entails some degree of heterogeneity and pluralism (Eisenstadt and Shachar 1987). No matter how successful an élite is in establishing and legitimizing common norms, these are unlikely to be fully accepted by all the members of society. Thus in any social order there is always an element of dissent. Even in the imperial cities of Beijing, Seoul, Tokyo, and Hanoi, minority subcultures existed throughout history.

The inevitable development of opposing forces within a society brings with it not only social and political change, but also the transformation of urban space. This interaction of society and space is occasionally marked by dramatic transitions that signal the birth of a new epoch. Examples from East Asia include the construction of Edo when the authority of the Tokugawa *bakufu* (the shogunate government) was solidified in the early seventeenth century, the construction of colonial enclaves in China, Vietnam, and Singapore in the nineteenth century, and the capture and reconstruction of Beijing by communist forces in 1947. More often, however, change comes gradually through the daily politics of appropriating, occupying, and developing urban space. In either case, the built environment is not just a backdrop for political and social drama but is an integral part of it.

Potential opposition groups often originate as secondary, disaffected élites (Vogel 1979). This was the case during the Meiji Restoration in Japan, which was led by lower-ranking *samurai*. Opposition groups may occasionally unite their forces and increase their strength by claiming sites or elements of the built environment that represent continuity with a past cultural order. The Communist Party of China's choice of Beijing as their national capital and of Tiananmen Square as the site for proclaiming the People's Republic in 1949 represented powerful uses of symbolism to capture popular support for restoring China's greatness (Samuels and Samuels 1989).

History is replete with evidence that the breakdown of integrative and regulative mechanisms is inherent in every society. Every type of political and economic system delineates specific operational boundaries; yet the very construction of such boundaries generates conflicts and contradictions that may lead to accommodation, transformation, or decline (Eisenstadt and Shachar 1987). This tendency toward opposition and change almost invariably focuses on the city. The Meiji Restoration, which moved the emperor from Kyoto to the renamed capital of Tokyo, is but one example of the association between social change and the city, both as a locus of control and as a symbolic centre of moral order.

The very conditions that give rise to a new social order invariably generate new social and political conflicts within the city. The establishment of a new orthodoxy is never total: the very forces leading to a new hegemony may themselves generate patterns that work against the orthodoxy. The concentration of population in cities and the dynamics of urban life in themselves create social and physical spaces that may work against political order. However, cities also contain the possibility of transforming emerging forces into a new orthodoxy. This pivotal feature of the dynamics between culture and the city is directly linked to a central problem of urban analysis: whether cities merely reflect or manifest social forces on a physical and spatial plane or whether they are also a force in themselves for social continuity or change. The approach taken here, while stressing the importance of the larger social, political, and cultural context (including outside forces), builds on the latter perspective of the city—as a built environment and part of a larger spatial process of appropriation and concentration of power—as a causal force and not merely as a reflection of culture.

The design and construction of cities provide a necessary physical and symbolic basis for any complex social order as well as foundation for social and cultural change (Abu-Lughod 1991). As Paul Wheatley (1972) stresses, the spatial concentration of cities generates new forms of production, control, and diffusion of information—technical, instrumental information as well as the basic symbolic universe shared by members of a society. Urban concentration also leads to modes of resource control that create a potential for new types of social and cultural creativity.

Function, Symbolism, and Power

The organization of any society relies on a concomitant ordering of space. This ordering is functional, in providing the physical requirements for the society to sustain and reproduce itself, and symbolic, reflecting cultural and political meanings and power in the built environment. Social changes invariably set in motion processes of spatial restructuring to support the new order. In general, the greater the change in social, cultural, political, and economic relations, the more profound the recasting of urban space.

The culture of a society is manifested in the built environment of a city in at least three major ways. The city is, first, the spatial counterpart of underlying economic organizations, be they markets or non-market institutions of economic exchange. Second, the design of a city and its various components represents moral codes and aesthetic rules that bring culture and function into the built environment. Third, the city reflects principles of spatial organization oriented toward maintaining security and political dominance. The concrete form of a city is influenced by all these considerations; any city is a mosaic reflecting their combinations and relative strengths in a specific geographical setting (Cohen 1976).

Throughout East Asia specific urban sites have been subject to overt symbolic manipulation by both rulers and the ruled. Tiananmen Square in

Beijing, Kwanghwamoon in Seoul, and Hibiya Park in Tokyo are among the more obvious examples. Beijing, the Forbidden City, has long been a symbol of imperial power and its accompanying cultural ideology. The invisible power of the imperial system is also symbolized in the heart of Tokyo. Such symbolism serves the purposes of social integration and regulation, often helping to maintain the hegemony of dominant political élites.

In the more overtly commercial culture of Hong Kong, the urban land-scape is a manifestation of the functioning and also the ideology of market capitalism, implanted and guarded by British colonialism. The passion for profit provides a major impetus for urban growth and redevelopment and displays its opulence in skyscrapers and luxurious hillside residences.

As these examples suggest, the role of the state cannot be ignored in a discussion of urban transformations in East Asia. The historical emergence of sovereign political entities exercising control over a territorial domain was inseparable from the rise of the city as a centre for the concentration and appropriation of resources. This emergence was not accomplished without cultural symbolism written into the built environment: city-building was part of a process of fortification, containment, and the direction of access to power and consumption.

The remarkable economic growth of countries such as Japan, Korea, Hong Kong, Singapore, and Taiwan has revived interest in the role of the state in economic development (Amsden 1989; Wade 1990; Haggard 1990; Vogel 1991; Douglass 1994). In socialist countries such as China and Viet-nam, the state has also played a powerful role in all areas of social life (Forbes and Thrift 1987). Urbanization and industrialization are obviously interconnected, and building a base for export-oriented manufacturing has necessitated an unprecedented expansion of state activity in city-building and management. This has included not only the provision of basic infra-structure for manufacturing, but equally important state intervention in addressing the needs of labour. Thus while governments have become engaged in the location of industries and the redevelopment of central urban areas for commercial use, they have also become heavily involved in public and subsidized housing—even in a so-called *laissez-faire* economy such as Hong Kong's. This has radically changed the built environment of cities such as Seoul, Hong Kong, and Singapore.

States—and those holding state power—have also become involved in urban development to improve their financial position. For example, the Japanese government's fiscal crisis in the late 1970s led to the formation of highly profitable state–business partnerships in urban development (Machimura 1992). This state–business collaboration in granting develop-ment permits to the private sector in exchange for official and unofficial financial returns effectively deprived ordinary citizens of land, housing, and a healthy, pleasant urban environment (Watanabe 1992). This led to considerable social tension (Douglass 1993). The situation was not much different in the late 1980s in Seoul.

Since their adoption of the market system, the socialist cities of Beijing and Hanoi have had to face similar tensions between groups competing for access to urban land and housing. Their urban landscapes have been changing quickly as modern office buildings, high-rise apartments, and hotels dwarf pre-colonial and colonial buildings and neighbourhood designs, indicating changes in social structure as well as architecture.

Cultural values and institutions have been deeply affected by the process of industrialization and urbanization. While some see this as a positive sign of movement toward 'modernity', others decry the changes taking place. Indeed, the perceived 'moral breakdown' in Vietnam after market reforms has prompted some to call for a revival of ancient Confucian values that stress social harmony and the importance of family. Some believe that an updated version of Confucianism could serve as a moral counterbalance to the rapid economic and social changes sweeping the country (Hiebert 1994). The current tension between old and new provides a backdrop for the project to preserve the Thirty-Six Ancient Streets quarter in Hanoi.

Such tensions are also illustrated in the rapidly changing coastal cities of China, especially in Shanghai. Here local leaders hope to attract financial houses and industries looking to escape from the soaring land and labour costs of Hong Kong. Yet this period of rapid social, political, and economic change poses an identity crisis. Many worry that the current passion for economic growth could jeopardize the balanced development of the city and even threaten its ability to attract much-needed foreign capital (Sender 1994).

The City in a New Global Order

The modern history of East Asian cities has been strongly affected by a complex web of local forces and international influences arising from the development of global capitalism. New transportation and communication technologies have brought new cultural values into societies that, in the past, enjoyed remarkably long periods of isolation. These new values are often highly materialistic and oriented toward the individual.

The power of global forces seems so pervasive that cities in East Asia appear to some observers to be converging with their counterparts in the West (Agnew and Duncan 1989; King 1989). Despite national efforts to maintain cultural and ideological independence, foreign travel, overseas study, and modern media have infused East Asian populations with a taste for the latest consumer goods and other life-style of advanced market economies. At the same time, the diffusion of production technologies increases the similarities in consumer and industrial goods (Behrman and Rondinelli 1992).

While the influence of global capitalism is pervasive, the structure and dynamic of each society is unique. Both global and local forces shape urban life and the urban landscape (Storper and Walker 1989). Each society encounters a juncture in history calling for substantial social, political, and economic restructuring to renew the basis for economic growth and ever-increasing levels of material welfare. The city is a key element in this process, both as the built environment for production and consumption and as the arena for the development of a more inclusive political process.

As the setting for production and consumption by urban populations, cities in East Asia are threatened by severe environmental degradation and pollution, affecting human health, well-being, and productivity (UNES-CAP 1993). Such problems are likely to become a central issue for urban middle classes as they make new demands for better living standards. With an expansion of this group and a rise in social awareness, there will be continuing demands for political reform and greater accountability in the state's use of power and public resources. In more general terms, there is a call for rethinking and restructuring the relationship between society and the state. This could take the form of weakening state control, as in South Korea, or maintenance of a strong state supported by a communitarian political culture, as in Singapore. Two things are certain, however: the pressure for political and economic restructuring is increasing in East Asia, and the work of restructuring will largely take place in the region's cities.

Economic restructuring is already apparent in Japan, South Korea, Hong Kong, and Singapore. The relocation of labour-intensive production from these countries to lower-cost environments in other parts of Asia entails not only enterprise restructuring but also spatial transformation (Clark and Kim 1995) and the development of new economic enterprises in the high-cost cities that are losing their industrial base. The information industry has been touted as the key to future growth. The development of an information industry, however, is closely linked not only with the level of technology in a particular city, but also with local culture in a broader sense. Murakami (1992) predicts that the production of 'consummatory' information (including knowledge, art, and language communication) will eventually replace the production of 'instrumental' information. In this event, a city's spiritual and cultural life will have a profound influence on its future development.

The trend toward deregulation and privatization in East Asia, in particular the waning of the national government's role in domestic markets, means that local governments and private enterprises are assuming more decision-making authority (Kim 1993; Yuen and Wagner 1991). In East Asia, local governments are making deals directly with transnational capitalists and foreign commercial interests. The trend toward deregulation and privatization could mean, however, the withdrawal of the public sector from the provision of collective consumption goods, which could result in greater social cleavages and deteriorating urban environments. In this context, economic and political reforms should not be confused with a diminished scope for action in the public domain. In many instances, especially in cities, government action needs to be enlarged in several areas. In East Asia, rapid economic growth and social change have also led to changes in social structures and values, widening economic inequalities, and increasing environmental degradation—all problems that are not easily amenable to market solutions (Douglass 1990).

As cities in East Asia become more closely enmeshed within the global community, there is a strong urge to retain or recapture each city's unique cultural heritage as a basis for urban identity and renewed creativity (Tu 1991; Sakakibara 1993; Wilson and Dirlik 1994). In this context, the rich traditions of East Asian cities need to be re-examined. The wholesale adoption of urban concepts from the West raises the spectre of social, political, and economic malaise widely reported from Western cities (Harvey 1992; Gans 1993; Marcuse 1993). In the context of global environmental degradation, the virtues of ancient Asian philosophies, such as harmony among persons and with nature, are worthy of being reappraised. Certain social practices derived from Confucian values, which have contributed to the rapid economic development of these countries and influenced the pattern and structure of cities, need to be analysed for a full understanding of cultural impact. It is, however, cautioned that Confucianism, like any other social philosophy or system of values, cannot be treated as timeless or ahistorical, and Confucian traditions vary widely among the societies of East Asia. Nevertheless, the essentials of Confucian philosophy—in combination with selected Western values—may provide alternative models of development. Such philosophical underpinnings may help bring back cultural and ethical dimensions to the discussion of the future of Asian cities.

The three major themes of 'culture and urban transformation', 'function, symbolism, and power', and 'the city in a new global order', which we set out in this introduction, are addressed in separate parts. Chapters in Part I deal with urban and urbanization processes at the macro level, focusing on the interrelationship between these processes and large social changes. With variations in their emphasis on the cultural and the political economy dimension of urban transformation, authors of case studies in Part II examine the relationship between the built environment and cultural, socio-economic, and political processes. To a certain extent, their analysis corresponds to our second theme: the function, symbolism, and power. The final chapter expands on the theme of the city in a new global order and suggests the potential role of culture in shaping the future of East Asian cities in the twenty-first century.

References

Abu-Lughod, Janet (1991). *Changing Cities*. New York: Harper Collins.
Agnew, John A., and Duncan, James (1989). Introduction, in John Agnew and James Duncan (eds.), *The Power of Place*. Boston: Unwin Hyman.
——Mercer, John, and Sopher, David E. (1984). *The City in Cultural Context*. Boston: Allen & Unwin.
Amsden, Alice (1989). *Asia's Next Giant: South Korea and Late Industrialization*. New York: Oxford University Press.
Basu, D. (ed.) (1985). *The Rise and Growth of Colonial Port Cities in Asia*. Lanham, Md.: University Press of America.
Beg, Muhammad A. J. (1986). *Historic Cities of Asia*. Kuala Lumpur: Percetaken Ban Huat Seng.
Behrman, Jack N., and Rondinelli, Dennis A. (1992). 'The Cultural Imperatives of Globalization: Urban Economic Growth in the 21st Century', *Economic Development Quarterly*, 6: 115–26.
Berger, Peter L., and Hsiao, Hsing-Huang (1988). *In Search of an East Asian Development Mode*. New Brunswick, NJ: Transaction Books.
Castells, Manuel (1977). *The Urban Question*. London: Edward Arnold.
——(1983). *The City and the Grass Roots: A Cross-cultural Theory of Urban Social Movements*. Berkeley and Los Angeles: University of California Press.
Clark, Gordon, and Kim, Won Bae (1995). *Asian NIES and the Global Economy: Industrial Restructuring and Corporate Strategy in the 1990s*. Baltimore: Johns Hopkins University Press.
Cohen, E. (1976). 'Environmental Orientations: A Multidimensional Approach to Social Ecology', *Current Anthropology*, 17: 49–69.
Costa, Frank J., Dutt, Ashok K., Ma, Laurence J. C., and Noble, Allen G. (1989). *Urbanization in Asia*. Honolulu: University of Hawaii Press.
Douglass, Mike (1990). 'Structural Change and Urbanization in Indonesia: From the "Old" to the "New" International Division of Labor'. Discussion Paper 15, Department of Urban and Regional Planning, University of Hawaii, Honolulu.
——(1993). 'The "New" Tokyo Story: Restructuring Space and the Struggle for Place in a World City', in K. Fujita and R. C. Hill (eds.), *Japanese Cities in the Global Economy: Global Restructuring and Urban-Industrial Change*. Philadelphia: Temple University Press.
——(1994). 'The "Developmental State" and the Asian Newly Industrialized Economies', *Environment and Planning*, 26 (A): 543–66.
Eisenstadt, S. N., and Shachar, A. (1987). *Society, Culture, and Urbanization*. Newbury Park, Calif.: Sage Publications.
Forbes, Dean, and Thrift, Nigel (eds.) (1987). *The Socialist Third World*. New York: Basil Blackwell.
Fuchs, Roland J., Jones, Gavin W., and Pernia, Ernesto M. (1987). *Urbanization and Urban Policies in Pacific Asia*. Boulder, Colo.: Westview Press.
Gans, Herbert J. (1993). 'From "Underclass" to "Undercaste": Some Observations about the Future of the Postindustrial Economy and its Major Victims', *International Journal of Urban and Regional Research*, 17: 327–35.

Ginsburg, Norton, Koppel, Bruce, and McGee, T. G. (eds.) (1991). *The Extended Metropolis: Settlement Transition in Asia.* Honolulu: University of Hawaii Press.

Goldstein, Sidney, and Sly, David D. (eds.) (1975). *The Measurement of Urbanization and Projection of Urban Population.* Liège: International Union for the Scientific Study of Population.

Haggard, Stephan (1990). *Pathways from the Periphery: The Politics of Growth in the Newly Industrializing Countries.* Ithaca, NY: Cornell University Press.

Harvey, David (1992). 'Social Justice, Postmodernism, and the City', *International Journal of Urban and Regional Research*, 16: 588–601.

Hauser, Philip M., Suits, Daniel B., and Ogawa, Naohiro (1985). *Urbanization and Migration in ASEAN Development.* Tokyo: National Institute for Research Advancement.

Hiebert, Murray (1994). 'Uneasy Riders', *Far Eastern Economic Review*, 157 (35): 54–5.

Jones, Gavin W. (1983). 'Structural Change and Prospects for Urbanization in Asian Countries', Papers of the East–West Population Institute 88. Honolulu: East–West Center.

Kim, Won Bae (1993). 'Economic Restructuring and Urbanization in Asia', in *Proceedings of the Nihon University International Symposium on Trends and Prospects of World Urbanization.* Tokyo: University Research Center, Nihon University.

King, Anthony D. (1989). 'Colonialism, Urbanism, and the Capitalist World Economy', *International Journal of Urban and Regional Research*, 13: 1–18.

Lawrence, D. L., and Low, S. M. (1990). 'The Built Environment and Spatial Form', *Annual Review of Anthropology*, 19: 453–505.

Lefebvre, Henri (1991). *Production de l'espace* (The production of space). Trans. Donald Nicholson-Smith. Cambridge, Mass.: Blackwell.

McGee, Terry G. (1967). *The Southeast Asian City.* London: Bell.

Machimura, Takashi (1992). 'The Urban Restructuring Process in Tokyo in the 1980s: Transforming Tokyo into a World City', *International Journal of Urban and Regional Research*, 16: 114–28.

Marcuse, Peter (1993). 'What's so New about Divided Cities?', *International Journal of Urban and Regional Research*, 17: 355–65.

Murakami, Yasususke (1992). *Han-koten no seijikeizaigaku* (The anti-classic political economy). Tokyo: Chuokoronsha.

Pye, Lucian (1985). *Asian Power and Politics.* Cambridge, Mass.: Belknap Press.

Redfield, R., and Singer, M. S. (1954). 'The Cultural Role of Cities', *Economic Development and Cultural Change*, 3: 53–73.

Richardson, Harry W. (1990). 'Urban Development Issues in the Pacific Rim', *Review of Urban and Regional Development Studies*, 2: 44–63.

Rondinelli, Dennis A. (1991). 'Asian Urban Development Policies in the 1990s: From Growth Control to Urban Diffusion', *World Development*, 19: 791–803.

Ross, R., and Telkamp, G. (eds.) (1985). *Colonial Cities.* Dordrecht: Martinus Nijhoff.

Rotenberg, Robert (1993). Introduction, in Robert Rotenberg and Gary McDonogh (eds.), *The Cultural Meaning of Urban Space.* Westport, Conn.: Bergin & Garvey.

Sakakibara, Eisuke (1993). *Bunmei toshideno nihon Nihon-gata Shihonshugi* (Japanese-style capitalism as a civilization). Tokyo: Toyokeizai Shimposha.

Samuels, Marwyn S., and Samuels, C. M. (1989). 'Beijing and the Power of Place in Modern China', in John A. Agnew and James Duncan (eds.), *The Power of Place*. Boston: Unwin Hyman.

Sender, Henny (1994). 'Passion for Profit', *Far Eastern Economic Review*, 157 (25): 54–6.

Smith, D. A., and Nemeth, R. J. (1986). 'Urban Development in Southeast Asia: A Historical Structural Analysis', in D. Drakakis-Smith (ed.), *Urbanization in the Development World*. Beckenham: Croom Helm.

Soja, E. (1989). *Postmodern Geographies*. London: Verso.

Storper, Michael, and Walker, Richard (1989). *The Capitalist Imperative: Territory, Technology, and Industrial Growth*. New York: Basil Blackwell.

Tu, Wei-ming (1991). 'Cultural China: The Periphery as the Center', *Daedalus*, 120: 1–32.

UNESCAP (United Nations Economic and Social Commission for Asia and the Pacific) (1993). *State of Urbanization in Asia and the Pacific, 1993*. Bangkok: UNESCAP.

Vogel, Ezra (1979). 'Nation-Building in Modern East Asia: Early Meiji (1868–1890) and Mao's China (1949–1971)', in Albert M. Craig (ed.), *Japan: A Comprehensive View*. Princeton: Princeton University Press.

——(1991). *The Four Little Dragons: The Spread of Industrialization in East Asia*. Cambridge, Mass.: Harvard University Press.

Wade, Robert (1990). *Governing the Market: Economic Theory and the Role of Government in East Asian Industrialization*. Princeton: Princeton University Press.

Watanabe, Yozo (1992). 'The New Phase of Japan's Land, Housing, and Pollution Problems', *Japanese Economic Studies*, 20: 30–68.

Wheatley, P. (1972). 'The Concept of Urbanism', in P. J. Ucko, R. Tringhan, and G. W. Dimbleby (eds.), *Settlement and Urbanism*. London: Duckworth.

Wilson, Rob, and Dirlik, Arif (1994). 'Introduction: Asia/Pacific as Space of Cultural Production', *Boundary*, 2 (21)(special issue): 1–14.

Yuen, Ng Chee, and Wagner, Nobert (1991). 'Introduction: Marketization of Public Enterprises', in Ng Yuen and Nobert Wagner (eds.), *Marketization in ASEAN*. Singapore: Institute of Southeast Asian Studies.

PART I

Culture, Society, and the City in East Asia

2

Culture, History, and the City in East Asia

Won Bae Kim

> Western man has been investing his efforts and his treasure in the mastery of his material environment. Now that his achievement of this mastery is forcing him back upon himself in self-defence against the dehumanised world that his technology has conjured up, he finds himself at a loss; and the non-Western majority of the human race is going to fall into the same straits as it becomes more and more deeply implicated in the 'extrovert' modern Western way of life. Megalopolis is going to swallow up human beings of all cultures, religions and races; for all of us, the problem of having to live in Megalopolis will have to be solved in spiritual terms.
>
> (Toynbee 1967: 27–8)

Cities in East Asia have been undergoing critical transformation since the Western encroachment in the nineteenth century. The penetration of capitalism and rapid industrialization in the twentieth century resulted in a tremendous concentration of population and industry in large cities of East Asia. Tokyo, now approaching a population of 30 million, has become an engine of growth for Japan and perhaps for the Asia-Pacific region. Despite its elaborate networks of transportation and communication, Tokyo occasionally encounters functional lapses. The social order, nurtured by the *samurai* culture, sustains the balance between the superstructure and infrastructure of the city.

Seoul is not far behind Tokyo. With its population of around 15 million, the Seoul metropolitan area produces almost half of South Korea's gross national product. Perhaps more conspicuously than Tokyo, the city is characterized by displays of lavish consumerism, both along the main boulevard and in the back alleys. The traffic jams appear insoluble.

After a period of hiatus, Beijing's skyline is finally changing. The Forbidden City is rapidly being enclosed by the rising walls of hotels, offices, and condominiums. Privatization of the planned economy adds new colour. The solidarity claimed by socialism fast disappears along the newly built highway between the city and the airport, with walls erected between the haves and the have-nots. There is a floating population of at least 1 million in addition to the official estimate of 10 million permanent residents.

Hanoi is now turning a new corner. With strong national pride, the citizens now also crave recognition from the international community. With foreign investment, construction is booming throughout Hanoi, and the city risks losing its urban identity. Unregulated development by private interests relying on foreign capital may make Hanoi a second Bangkok.

After a remarkable accumulation of wealth on borrowed time, Hong Kong now faces reversion back to China. Will this city's unashamed money-making be too much for the Chinese, or will Hong Kong provide a model for the rest of China?

Singapore strives for a central position in South-East Asia, with a state-sponsored scheme to develop Singapore with Johor, Malaysia, and Riau, Indonesia, as a growth triangle. Through elaborate social engineering, Singapore seeks to establish a new identity, not merely as an economic centre but as a viable cultural centre combining British vision with Chinese morality.

Where are these great cities of East Asia going? Admitting that there are significant differences in the way each of them responds to internal and external pressures of change, at a general level they may share concepts and approaches that separate them from the Western experience. Are they evolving as distinct East Asian models, and, if so, what distinguishes them from cities in the West? Culture and history certainly account for the important differences in ideology, power structure, and economic system of each society.

This chapter surveys the structure and evolution of Asian cities in the historical and cultural context. The main purpose is to provide a perspective, lacking in contemporary discussion of urban development and planning, by properly situating cities in their larger social, economic, and political context. A city is understood here as the embodiment of a society: an old city is the embodiment of the society's history and culture, with multiple layers going back over time. This perspective facilitates our understanding of urban change over time and helps us participate in the future of our cities in a meaningful way. The discussion will begin with the conception and design of the city in East Asian culture. Second, dynamics of urban transformation will be discussed in relation to underlying socio-political processes by employing the concepts of centrality and concentration developed by Eisenstadt and Shachar (1987). Third, the discussion will focus on the interplay of global forces and local responses in the transformation of the urban landscape. Finally, the social and cultural creativity of East Asian cities will be discussed in the context of increasing local–global interactions.

The Traditional City in East Asia

In the West, cities have often been defined or identified by contrasting them with non-cities. The antithetical relationship between the city and the non-city was largely absent in East Asia, particularly in Confucian societies such as China, Japan, and Korea. As Smith (1979) discusses, the English concept of the 'country' as pastoral was not necessarily antithetical to the concept of the city. The Romantics in nineteenth-century England reinterpreted the concept of nature, which came to signify not an underlying order of which man was an integral part, but rather a separate order untouched by man. In contrast, East Asian traditions, at least in Taoist and Buddhist societies, assumed a natural order of which man was an integral part. In Japan, the city was conceived as a place of mediation between humankind and nature, affirming a simultaneous multiplicity rather than a dichotomy between commercial and agrarian functions (Smith 1978).

Similarly, East Asian conceptions of 'country' are not inherently anti-urban. The idea that the city represents a higher level of civilization than the countryside is a cliché of the Western cultural tradition. Chinese civilization did not separate the urban from the rural. Rather, an urban–rural continuum was possible through practical access to both geographical and social mobility (Mote 1977). The close interdependence of city and countryside was explicitly recognized (Murphey 1984). These concepts were basically temporal rather than spatial, as reflected in Japanese words such as *furusato* (hometown) and *inaka* (countryside) (Smith 1979). The Confucian belief that agriculture is the base of a properly functional society (*nongbenzhuyi* in Chinese) is apparent in East Asian rice cultures that stress the moral worth of an agrarian way of life. The dominant notions of 'city' and 'countryside' in East Asia are based on a social, rather than on an aesthetic, order.

The city or town in East Asia represents a social order created by state 'action' (*zuowei* in Lao Zi philosophy) in contrast to the 'natural order' (*ziran zhixu*). The city is contrasted with the village, which is a natural, spontaneous settlement based on kinship and territoriality. This concept of the city is deeply rooted in Chinese philosophy, where the dual category of *yin* and *yang* represents two inseparable opposing forces governing the universe. The dialectic of *yin* and *yang* applies all the way from cosmic order down to daily life, affecting every aspect of human life. Even the site of a person's grave is carefully selected according to the rules governed by innate forces in nature. The city, which represents *zuowei*, emulates the natural order, from its site selection to its spatial arrangement. Thus, the city is more than a political idea; it is a metaphysical, transcendent concept, as demonstrated in the original layout of traditional cities in East and South-East Asia.

In East Asia, the origin of the concept of the city is inseparable from the history of institutions, which is none other than the way of defining human beings and their relationship with nature and other men (Kim 1991). Early Chinese philosophical thought was primarily concerned with the relationship between humankind and nature: the relationship between human beings and the construction of an ideal community became the major focus of Confucian philosophy. Distinctively Chinese social relations were manifest in the built environment of Beijing, whose essence is represented by walls. 'Over and above practical considerations, the wall is the line clearly drawn between what is significant and what is insignificant, what is powerful and what is not powerful, who is kin and who is stranger, what is sacred and not sacred. The Chinese passion for walls reflects their passion for clarity in human relations, signifying an individual's identity and place within society' (Meyer 1991: 4).

The ideal state or community, as conceived by Lao Zi, has a small population that does not need to use tools, weapons, ships, or carts, even though it has them all. People are content with their life in nature and are not bothered by dealings with people of other states (Lau 1981; Chang and Lee 1988). As non-effort (*wuwei*) is considered the most desirable state of human life in Lao Zi philosophy, a community with minimal state intervention is regarded as an ideal place to live.

While employing certain elements of Taoist thought, such as harmony and balance, Confucian philosophy constructed an ideological basis for central statism, which emphasized hierarchy and order. More importantly, social relations, including interpersonal relations, were central to the Confucian ideology, which had a decisive impact on the internal structure of cities in East Asia. The development of centralized bureaucracies and the concomitant elaboration of socio-political order were closely paralleled by the rise and spread of urbanization in East Asia (Chang 1963). The establishment of *xian* capitals in frontier regions marked the expansion and penetration of central authority and the creation of social order. New cities were not self-governing, but rather grew up as administrative units of the central government (Eberhard 1956).[1] This contrasted with the relatively independent development of cities in the West during the Middle Ages. Note that cities are becoming independent again in a borderless economy, as in the European Union. The degree of centralization, however, differed among the countries of East Asia. For example, in Tokugawa Japan, cities and towns had some autonomous power from higher political authority such as the *bakufu* (the shogunate government) and the emperor, although their autonomy was less complete than that of cities in Europe.

---

[1] Rozman argues that the principal force behind the development of an urban network in pre-modern China was administrative centralization, while commerce remained secondary (Rozman 1978).

The Design of Traditional Cities

Present-day Beijing was founded in 1267 during the Yuan dynasty. The design of the city of Dadu was similar to an ideal design of imperial capitals first published in a book entitled *Zhouli Kaogongji* (A study of engineering) in the fifth century BC. The overall design of Beijing closely followed two basic, overlapping liturgies: (1) the cult of the emperor, whose cosmology defined him as the Son of Heaven and his place as the axis of the earth, and (2) an arcane, partly shamanist, partly Taoist, Confucian-assimilated geomancy (*fengshui*: the Chinese art of placement) based on such ritual and divination texts as the *Zhou Li* and the *Yi Jing*. These two traditions combined to establish the basic design principles of centrality, axiality, and rectilinearity (Samuels and Samuels 1989).

Hence, the order impressed on Changan is the cosmic order, a pattern derived from the motions and positions of the heavenly bodies, which, if realized on earth, would ensure the stability and permanence of the capital, realm, and dynasty (Meyer 1991). Meyer (1991) groups some of the most general features of these sacred cities into four characteristics: (1) orientation, the attempt to lay out a city in accord with the geometry of the cosmos; (2) centrality, with the city itself called the centre of the world or, in other cases, a central place identified as a well that reaches down to the underworld, a cosmic mountain, the temple of the god, or a royal palace; (3) association with a sacred king who is the channel of divine power, the conduit through whom the blessings and warnings of the divine beings are brought to earth; and (4) division into a stratified social order with clear differentiation of functions. The cities are an expression of hierarchy in concrete form. The social order and urban structure seem to mirror the same underlying reality—the organizational conception behind the creation of city and society. These cities are clear geometrical forms, not just the result of spontaneous growth or agglomeration. They are 'planned' in structure and in function, following the divine pattern of government.

Having largely been under Chinese influence until the nineteenth century, urban structures in pre-modern Korea, Japan, and Vietnam followed the Chinese idea of city-building, but with variations. Kyongjoo in the Sillah dynasty, Seoul during the Yi dynasty, Nara and Kyoto in pre-Tokugawa Japan, and Hanoi under the mandarin monarchy (starting from the eleventh century) all adopted the cosmic order as the principle for city construction.

The structure of Beijing made clear to all levels of society, in a very concrete way, their place in the cosmic and social hierarchy. If urban residence did not bring freedom as it did in the mercantile cities of medieval Europe, it at least provided a kind of security and a measure of peace in the acceptance of one's assigned role. Chinese cosmology set specific

requirements for the organization of physical space. As a direct expression
of social and political structures, the built environment expressed or re-
affirmed through symbolic associations the relationship between groups
and positions held by individuals (Lawrence and Low 1990). As symbols,
sites such as Tiananmen Square in Beijing expressed powerful meanings
and values: they comprised key elements in a system of communication
used to articulate social relations. As will be discussed later, the complex
levels of meaning associated with such sites were manipulated by political
actors for a variety of purposes in different situations. The arrangement of
sites thus ultimately corresponded to the structure of society (Lawrence and
Low 1990).

Dynamics of Urban Transformation in East Asia

Following Eisenstadt and Shachar (1987), we argue that explaining
urban transformation by reference to either 'culture', 'mode of production',
or 'political economy' is too simplistic (Sawyer 1975; Walton 1984). Rather,
we propose to look at urban transformation in the larger political, social,
and cultural context using a historical perspective centring upon the critical
role of the economic system (King 1984; Tilley 1984). This proposition
assumes that the continuous interaction between the forces shaping social
division of labour, the regulation of power, and the construction of trust and
meaning sets the contours of institutional structures, particularly the structure of cities and urban systems. These interaction, however, take place in
different political-sociological settings, which are influenced by the structure of the international political and economic system, the ecological features of the city, and the city's dependence on internal and external
markets.

The urban transition in East Asia, perhaps during its entire history, can
be explained by two key concepts. These are concentration and centrality
(Eisenstadt and Shachar 1987).[2] Concentration refers to the agglomeration
of population as a result of demographic and economic processes, which in
turn generate processes of social differentiation and division of labour.
Centrality refers to the process of constructing the cultural, political, and
moral order of a society in a specifically defined space and environmental
setting and the process of dominating society from the centre (Eisenstadt
and Shachar 1987). Concentration is a manifestation of material, economic
forces, while centrality represents spiritual, cultural forces.

[2] Concentration and centrality closely resemble Mumford's two principles: accumulation
and conquest (Mumford 1961). Also, concentration and centrality produce types of cities
parallel to Redfield and Singer's (1954) 'heterogenetic' and 'orthogenetic' categories. Concentration largely reflects economic forces, resulting in agglomeration economies that include
specialization and provide the mechanism to support urban growth (Hoover and Giarratani
1987). In contrast, centrality reflects the cultural, political, and moral order of a society.

Concentration and centrality cannot be discussed in isolation from the political and economic setting. Feudalism and centralism (with a strong central state) are the dual categories of the historical evolution of Asian society. Feudalism means, in a broad sense, a tendency towards decentralization. The pendulum swing between feudalism and centralism is evident in China's history. A strong tradition of centralized government in China and Korea is contrasted with a tradition of decentralization in Japan. Because of this difference, it is not surprising that the role of Edo (Tokyo) in Japan was much weaker in the pre-industrial era than of Beijing in China or of Seoul in Korea.

In practice, the two processes of concentration and centrality frequently overlap. Every society contains elements of both, in unique combinations. Hong Kong and Singapore were largely created by colonial powers and never played a role as political or spiritual centres comparable to that of Beijing, Seoul, Tokyo, or Hanoi, nevertheless they have certainly carried out a role as centres, both within and beyond their political boundaries.

The precise form a city takes in the process of concentration depends on the type of economy prevailing in the area, the level of development, and the mode of international economic relations. Along with concentration, economic and occupational diversification in a city determines its spatial differentiation. This process is distinct from the urban form generated by centrality, which reflects political, administrative, and religious functions. According to Eisenstadt and Shachar (1987), the evolution of urban form is influenced by three factors: (1) the interrelationship and relative importance of forces of concentration and forces of centrality; (2) the specific structure of the major social state, whether these forces are autonomous or embedded, monolithic or heterogenous; and their access to the centre; and (3) the size of the society and external relations. The interplay of concentration and centrality produces a specific spatial order and form. Of course, these forces are mediated through the political and economic structure of a society.

The simultaneous working of centrality and concentration often involves conflict and tension between the forces behind the two processes. The celestial Beijing has been challenged by heterogenetic forces, sometimes forcing reconfiguration of its centrality. Ironically, the invasion of Mongols and Manchus provided stimulus and thus helped sustain the centrality of Beijing and perhaps Chinese civilization itself. Beijing during the Qing era (1644–1911) contained heterogeneous groups or subcultures, divided by ethnic, religious, or other characteristics. This situation required an elaborate spatial organization to preserve social order (Dray-Novey 1993).

In Hanoi, a commoners' city has always coexisted with a royal citadel. Today the traditional and French colonial quarters are threatened by the construction boom led by foreign investment. In Edo during the late Tokugawa period, tension was apparent between the *shitamachi* ('low city' occupied by the commoners) and the *yamanote* ('higher ground' occupied by the *samurai*). The ensuing transformation of *yamanote* during the Meiji period, through the alignment of the imperial system with a newly formed class of merchants, redefined the centrality of Tokyo.

Political and Social Order in Pre-modern East Asia

The centrality of the state in social, economic, and cultural development has long been pervasive in East Asian societies. In contrast to other regions, the relationship between state and society in East Asia was sustained by a shared belief in a moral order, which gave the government legitimacy and the people security and peace. As Pye (1990) observed, in modern China there has been a frantic search for a new moral order to replace the eroding Confucian system instead of a drive toward pluralism and a system of legitimacy anchored in political realities. Similarly in Japan, Korea, Taiwan, and Vietnam, efforts have focused on establishing a society guided by a moralistic ideology. In all these countries, the intelligentsia played a central role in constructing and perfecting a Confucian moral order, sometimes driving their society to the brink of collapse as seen during the Yi dynasty and the Cultural Revolution in China. Despite occasional resistance to the government excesses, the predominant role of the state, supported by a Confucian ideology, largely crafted and refined by the intellectuals, was seldom challenged until the advance of Western imperialism in the nineteenth century. The rise of capitalist entrepreneurs and a middle class is a recent phenomenon that has accompanied the expansion of Western capitalism into East Asia: these groups are likely to play an increasing important role in East Asian societies in the future.

The political system in pre-modern East Asian societies had a centralized bureaucracy headed by a king at the apex of an aristocracy. This unitary political system had a decisive influence on the prevailing social order in most of East Asia, though to a lesser extent in Japan. The social order was hierarchical, and mobility was limited. Commoners were distinguished from mandarin in China, as they were from *yangban* (nobility) in Korea and *samurai* (warriors) in Japan. In China and Korea, the political and moral authority was concentrated in the king and the aristocracy. The political system in Tokugawa Japan (1615–1867) had a comparatively decentralized structure based on territorial organization of the *han* (feudal clan). Political authority was shared by the *han*, the *bakufu* (the shogunate government), and the emperor. The pluralistic pattern of political authority in Tokugawa Japan made it relatively easy for *samurai* to transfer their loyalty from *bakufu* to *han* to the emperor in a crisis situation. In contrast, the unitary system of political authority under Chinese and Korean dynasties was less conducive to reform. Instead, a rigid bureaucracy armed with neo-

Confucian orthodoxy was resistant to change (Sato 1979). This was particularly true in Korea.

In pre-modern East Asia, most resources were centralized by the state,
although there was some variation between countries. For example, the
class of landed *yangban* in Korea during the Yi dynasty was not hereditary,
and the members of this class were often challenged by the non-landed
yangban, whereas the *samurai* in Tokugawa Japan did not own land but
were guaranteed certain hereditary fiefs and stipends. The Korean clan
(*munjung*) was a kinship organization centred on the worship of common
ancestors, whereas the Japanese clan (*dozoku*) was a mutual assistance
group, significant only to those members who lived near each other. In
Korea, the *yangban* bureaucrats had no permanent guarantee of their positions so that the clan and clan-based factions served as an important source
of identity and support. In Japan, the *samurai* retainers were divorced from
the land: they were dependent on the *han* for their social status and income,
so their loyalty was directed primarily to the *han*. In this respect, the
samurai–han relationship resembles the employee–employer relationship
of more recent times. Unlike their Chinese and Korean counterparts, the
samurai were not strongly influenced by the anticommercialism of neo-
Confucian ethics (Sato 1979).

Under a strong central state, centre–local relations were asymmetrical
with localities dominated by the central power. This dominance was less
complete in China than in Korea and other countries because of its size.
The high primacy of capital cities such as Beijing and Seoul is a direct
reflection of the centralized bureaucracy in pre-modern times. In 1789
Seoul's population was 189,153, which was larger than the total population
of the next largest thirteen cities in Korea. In comparison, despite its large
size, Edo in Japan did not have as high a primacy rating as Hanyang (the
capital of Choson kingdom). The primacy in Japan was shared by the
triangle of Edo, Osaka, and Kyoto, reflecting the decentralized nature of
the political system.

The Confucian moral and political order discouraged the rise of industry
and commerce. Greed and self-interest were abhorred, while the virtues of
frugality and righteousness were encouraged. Korean society during the Yi
dynasty provides an example of extreme anticommercialism. In this context, the growth of a city was circumscribed by the extent to which it could
draw agricultural surplus from its hinterland. Social status, which was largely determined by moral and political ideology, reinforced the anticommerce
and anti-industry bias. Social and geographic mobility increased when the
ideology began to erode.

Indeed, the growth of Seoul from the mid-seventeenth century was largely due to the development of urban commerce and industry and concomitant in-migration from rural areas. By the late fourteenth century, the
ultraconservative interpretation of neo-Confucianism (principally the

Zhuxi school) had gone to the extreme of banning market places (*changsi*)
except in Seoul. The development of commercial cities was inhibited until
the late nineteenth century when the country was forced open by Western
forces. In Korea, the population of Hanyang remained at around 200,000
during the entire Yi dynasty, while in Japan Edo had grown to almost 1
million by the end of the Tokugawa period.

The strictly hierarchical social order pervasive in pre-industrial East
Asia was an impediment to social mobility, although a degree of fluidity
existed in China and Korea based on personal merit (Ho 1959). Individual
merit, rather than birth, was emphasized for the recruitment of the
ruling élites, at least in principle, but this principle operated differently in
each country. Japan in the Tokugawa period was much more rigid than
under the Ming (1368–1644) and Qing dynasties (1644–1911) or Korea
under the Yi dynasty. In Japan, the movement of individuals from one
status to another was not legally blocked. Although the examination system
allowed some upward mobility in Korea under the Yi dynasty, such
opportunity was limited largely to the upper and middle classes.[3] As
these traditional systems gave way to the forces of concentration, a substantial shift in the social composition of the population occurred. For
example, during the Yi dynasty, there was a significant increase in the
yangban stratum and an equally significant decline in the *sangin* (com-
moner) and *nobi* (servants) strata in Korea, which led to social and urban
transformation. This reversal was more pronounced in rural areas than in
the cities: in-migrants came to the cities to engage in commerce and industry, while a large number of *yangban* out-migrated from the cities to the
rural areas. This peculiar phenomenon stems from the role of the intellectuals in Confucian society. They acted as guardians of Confucian ideology,
engaging in educating children in the countryside. Intellectuals who did not
hold official positions had to fall back on the support of their clans, based in
local villages. The emergence of a regional culture during the mid-sixteenth
century was an important feature of the overall political landscape, which
also influenced the relationship between city and countryside (Deuchler
1992).

Social and Spatial Order within the City

The segregation of 'sacred' and 'mundane' was an inevitable result of
tension between these two spheres in Confucian societies. Compared with
South-East Asian cities, Chinese spatial concentration was characterized by

[3] This upward social mobility was important during the Ming and the Qing eras in China
(Ho 1959). Although less social mobility was observed during the Yi dynasty in Korea, more
than half of all *munkwa* (civil service examination) graduates between 1421 and 1630 came
from non-*yangban* families (Choe 1987).

greater differentiation between ritual and administrative functions and by a kind of organized gradation of commercial and residential districts according to their hierarchical position in the basic social structure (Eisenstadt and Shachar 1987). In Vietnam, the towns became larger and more important with the advent of the coming of the mandarin monarchy in the eleventh century. The urban structure was basically bipolar with the official, royal city at one pole, and a teeming commercial city at the other. Thang Long, the capital that became Hanoi in the fifteenth century, was a typical traditional city. The royal citadel occupied a small space in the middle of open countryside, whereas the commercial town demonstrated all the features of an urban agglomeration (Nhuan 1984).

Residential segregation by social status was evident in Hanyang, the capital city of the Yi dynasty. The area near the palace was occupied by the élite (so-called *pukchon* or northern town). The lower *yangban* and the *yangban* without government positions lived in the southern area of the city (*namchon*, southern town). Lower-ranking functionaries lived in front of government office buildings, which were concentrated in the central part of the city. Public markets were located along the two central east–west roads, forming a commercial district,with merchants and artisans living nearby. The lowest stratum of society lived outside the city wall.

A similar spatial arrangement was seen in Edo. The shogun and his family occupied the innermost compound, while favoured *daimyo* (feudal lords) and other high-ranking officials lived just outside the compound's main gate. Lower-level aristocrats were located further out from the centre, while servants and others without status lived outside the central complex. In addition to the fortified central area, the city consisted of two quite different districts, the *shitamachi* or 'low city' where the common people lived and the *yamanote* or 'higher ground' occupied by the *samurai* class (Cybriwsky 1991).

In the cities of China, Korea, and Vietnam, hierarchical social order also imposed strict size regulations on buildings and plots (Sohn 1977; Nhuan 1984). Even the style, patterns, and colours of houses were closely controlled to differentiate the social status of residents.

The Decline of Centrality

Even before the encroachment of Western power in East Asia, there were occasional resistance and reform movements against the rigid Confucian social order. The neglect of commerce, industry, and science was criticized by more pragmatic schools of thought. Despite suppression by imperial governments in China and Korea, indigenous commercial and industrial activities began to appear during the eighteenth and nineteenth centuries.[4] The invasion of Western colonial powers then forced a more radical transformation from the Confucian social order. The transition was largely toward a commercial order dictated by the first state of global capitalism (King 1989). For society, it meant the collapse of the centrality of the state and the hierarchical social order. For cities, it meant a transition from orthogenetic to heterogenetic patterns.

The story of Beijing vividly illustrates this transition: from 1860 (when the Yuanmingyuan Summer Palace was destroyed by the joint Anglo-French forces) to 1949 (when communist China was established), the whole fabric of Confucian China came apart at the seams and, along the way, the cosmology that had so long dominated the landscape of the celestial city fell into abandonment (Samuels and Samuels 1989). On 12 February 1912, with the formal abdication of the last emperor of China, the landscape of Beijing was officially desanctified. The first three decades of the twentieth century witnessed the wholesale transformation of Beijing's urban landscape, a process whereby the secular and the profane ultimately overran the sacred and the celestial fabric of old Beijing (Samuels and Samuels 1989).

The early twentieth century was characterized by urban social crises in China. Even though city élites and organizations had developed an impressive array of policies and strategies to cope with social conflict and dislocation in the centreless polity of China, no city or alliance of city-based forces ever proved capable of ordering the contradictory tendencies into a self-conscious, self-interested urban political order (Strand 1989).

The forced opening of Korea by the West also brought significant social changes. The urban structure of Seoul changed to accommodate the colonial and foreign powers, and Korea's relationship with China became less important. During the colonial years, Japanese investment and migration into Korea resulted in a dualistic spatial and economic structure in Seoul and other Korean cities. Japanese settlers concentrated in the cities and took high-wage jobs in the modern sector (Ko 1988), while indigenous capitalism was suppressed and the industrial structure was distorted (Kim 1985). The layout of Seoul was also partly reorganized, destroying the integrity of the city's original design. The construction of a colonial government building in front of the Kyongbok Palace in 1923 clearly demonstrated that the *chosenshotokufu* (Japanese colonial government) was the ruler (Yoshida 1992).

---

[4] Skinner (1977) argues that commercial growth in China started much earlier. There was an irreversible policy shift between the 8th and the 13th cents., and afterwards commerce outside the capital cities was never again suppressed nor a system of administrative controls introduced. In Japan, commerce was encouraged by competition among feudalistic powers (Chubu 1967). In Korea, state control of commerce and industry was weakened during the 17th cent., and the feudal commercial system was gradually replaced by an early form of modern commerce (Kang 1973).

This is also the period when colonial urban centres came into existence in other parts of East Asia, including Hong Kong, Singapore, Jakarta, and Saigon. These new cities, founded by the European officials during the mercantile era of Western capitalism, provided links between Asian and Western civilizations. Often located in coastal areas, they were indispensable to the colonizers for the export of Asian wares and raw materials and the import of manufactured goods from Europe (Beg 1986). Traditional cities were also subject to colonial influences as East Asian societies were colonized by Western powers or adapted to Western industrial patterns during the early twentieth century. Freed from the grip of centrality and a suffocating social order, cities in East Asia grew rapidly, largely following the process of concentration accompanied by material accumulation. Western capitalism brought with it new ideologies different from indigenous 'communitarian' values. The emergence of industrial capitalists and a middle class during rapid industrialization and urbanization in the second half of the twentieth century has significantly changed the urban landscape in East Asia. Even China and Vietnam, which pursued the construction of a socialist society, joined the capitalist order during the 1980s.

Despite recent developments, there remains a strong undercurrent of a moral order in the societies of East Asia and of centrality in their cities. Indeed, there are indications of a resurgent interest in centrality and indigenous moral systems. Whether the economic rise of East Asia will bring an 'inevitable' cultural clash with the West remains to be seen, but culture has become a primary focus of international politics since the end of the Cold War (Huntington 1993). The capital cities of East Asia are important sites for tensions and conflicts, and perhaps reconciliation. As nodes within an expanding global network, the cities present their inhabitants with a sharpened contrast between old and new and help raise their consciousness of their own culture and heritage.[5]

Urban Form and Cultural Continuity

The city can be conceived of as an idea that becomes visible as a physical form. Beijing is a prime example (Meyer 1991). Here, the idea of a city as a political and religious centre is still alive. In contrast, the cities in Western Europe show the persistence of physical remains despite the loss of their original planning concepts. The conception of Seoul, following the Chinese model, represents a modified version of a political centre, although lacking the religious function of Beijing. Tokyo was designed on a unique

---

[5] The nodal concept envisions the city as part of broad open systems that promote diversity and change, whereas the multiple time layer concept suggests cultural continuity in time and space (Spodek 1993).

indigenous model of a *jokamachi*, or castle town; these sprang up throughout Japan as power centres of local feudal lords.

Urban form, or the built environment of a city, reflects the political and cultural orientation prevalent in a society, especially among the political élites. This reflection may have been clearer in traditional cities than in modern ones, where economic forces combined with multiple actors to create a jumble. Centrality and concentration in combination create a social order manifested in the built environment. Residents of such a city perceive the social order throughout their daily lives. To most citizens in Beijing, Tiananmen Square is perceived not simply as an artefact but as a symbol of the old and new political order in China.

In China and Vietnam, a remarkable continuity seems to prevail in terms of spatial and political structure between the former mandarin regimes and the socialist regimes of today, despite the fact that the differentiation of space by social status has been removed to a certain extent according to the political ideology of egalitarianism (Nhuan 1984; Samuels and Samuels 1989). A fundamental task during the early years of the People's Republic of China was to transform the old capital, which had been planned and designed under the autocratic emperors, into the capital of a people who lived in a new era of socialism (Hou 1986). The siting of *renmindahuitang* (People's Great Hall) and Mao's mausoleum in Tiananmen Square reflected a manipulation of symbolism by the political authorities. Socialist planners in China justified the destruction of traditional buildings as symbols of feudalism. There are, however, walls around many facilities, factories, government office buildings, and residential complexes; what has changed is the identity of the people inside the walls (Meyer 1991).

The transformation of the urban landscape of Beijing to secular status, which has been on going since 1949, underscores the continuity of revolutionary desanctification. In the wake of such campaigns as the 1950s effort to transform former 'consumer cities' into modern 'producer cities', the proletarian ethos of socialist China has generated a new landscape, supplanting the forms of celestial China with the steel and concrete structures and saved networks of the industrial age (Samuels and Samuels 1989). Yet, although socialist, Beijing has none the less remained 'Chinese', symbolizing a variant of socialism derived from Chinese sensibilities. Amidst the disruption of the socialist transformation, and even the extraordinary iconoclasm of the Cultural Revolution, which attacked all vestiges of Confucian China, a 'renaissance' has emerged. Paradoxically, the ambiguities of the early twentieth century have been accompanied by powerful continuities in the People's Republic of China (Samuels and Samuels 1989). The conflict between the imperial, Confucian past and the socialist present has been portrayed in *He Shang* (The river eulogy), a six-part television series, which set the cultural agenda for an unprecedented national debate: 'Whither China?' (Tu 1992). More recently, a commercial order, led by capitalist interests, has emerged to dominate the urban landscape of Beijing. China's new capitalists will make a mark on Beijing just as profound as did their imperial and socialist predecessors (Huus 1994).

Much as in the Soviet Union and other so-called revolutionary nations, the historical continuity of certain places has remained undeniable (Cavalcanti 1992). Even as Beijing underwent all kinds of infrastructural and architectural transformations, the city has defied change and reasserted its historic role as a uniquely symbolic centre (Hou 1986; Samuels and Samuels 1989). Ironically, the leadership of the Chinese Communist Party not only launched the new People's Republic from the principal ceremonial gate of the Forbidden City but also chose part of its inner domain as the site for their own private residential and office complex, known as Zhongnanhai. Sited and designed as the symbolic heart of Beijing, Tiananmen Square became a powerful statement of the many ambiguities of modern China. Thus the built environment serves as a symbol of both ideology and power, as manipulated by the city's inhabitants and by the political authorities.

In state-led capitalist countries such as South Korea and Japan, historical continuity also remains strong. The emperor's palace occupies the heart of Tokyo, defying the forces of market capitalism. Indeed, the presence of the emperor and his palace in a strange, almost serene, void in the heart of Tokyo regularly reminds the Japanese of their traditional culture and moral order. In Seoul the remains of the Yi dynasty palaces are on public display, but the Confucian social and moral order espoused by the élites of the Yi dynasty does not live on in the minds of ordinary people in Seoul's bustling streets. Why have traditions not been nurtured in modern-day Korea? The recent debate on the removal of the Japanese colonial government building suggests that some feel the national spirit can be recovered by destroying the physical symbol of the thirty-six years of Japanese occupation despite the fact that the colonial legacy remains in less visible aspects of urban life.[6] In this context, it is interesting to note that the Japanese built this building intentionally at a critical site which blocked off the lifeline of the whole traditional palace complex. Since the Japanese occupation, Seoul has been subject to manipulation by successive political élites, all of whom have tended to be ignorant of or biased against Korea's past. After the disruption of the American occupation, the Korean War, and the division of Korea, post-war regimes have not paid much attention to the traditional culture or

---

[6] The colonial legacy is still discernible in modern Seoul, even though the discontinuity from the colonial past is also apparent. The industrial structure contains elements of both continuity and change: the concentration of small and medium-sized enterprises is partly related to the colonial pattern. Moreover, there seems to have been a certain continuity in urban planning and management until the early 1960s: the 1962 Redevelopment Plan contained the essence of the 1934 Urban District Plan devised by the Japanese. Until very recently, the authoritarian style of planning and development which was a legacy of the colonial period has been evident in Korea (Hashitani 1992).

moral order, apart from the selective application of Confucian values such as frugality and loyalty during the regime of Park Chung-Hee. The political élites have been busy legitimizing their power by relentlessly pursuing economic development. In the process, Koreans appear to have wanted to forget their past. This dissociation from history is in sharp contrast to Japan's careful nurturing of historical continuity.

Global–Local Dynamics and the Cultural Creativity of Cities

For East Asian countries, such as Korea during the Yi dynasty, Vietnam under the monarchy, and Tokugawa Japan, the Chinese world order provided the basic framework for international relations until it was challenged by the West. South-East Asia, by contrast, was exposed to the influence of both India and China. The strong identification in Korea with the Chinese Ming dynasty may even have been an impediment to the development of Korean nationalism (Sato 1979). Indeed, rigid adherence to the Chinese world order brought on the decline of the Yi dynasty in Korea and the institutional rigidities embedded in the social and cultural institutions led to the demise of the state in 1905. Vietnam was under the suzerainty of China until the encroachment of the Western powers and did not develop a strong sense of nationalism. In contrast, Japan retained a greater degree of political and cultural independence from China. These differences in political and cultural integration with China help explain the different responses of East and South-East Asian countries to the encroachment of the West during the eighteenth and nineteenth centuries.

The twentieth century saw a rapid political and economic integration of most East Asian societies into the Western industrial order. A reform movement initiated by *samurai* in the late nineteenth century made Japan the first country in the region to complete the course of industrialization and urbanization (Smith 1966). Today, Tokyo is truly a world city with a population of at least 30 million. China is generally expected to rise in international importance in the twenty-first century. If this happens, Beijing may reassert its centrality in the global political economy, and a new international order may emerge, shifting from a focus on the United States to a multipolar structure in which China and East Asia will feature prominently.

Over the past century or so, the dominant trend in East Asia cities has been concentration with the goal of economic growth. Some cities have succeeded in managing concentration by maintaining or constructing a cultural, political, and moral order around the concept of centrality while others have not. For example, the rise of Tokyo in the post-industrial era signifies the positive feedback between concentration and centrality, largely based on the city's superior technological capacity. Most of all, Tokyo's strengthened centrality has been possible because of a tradition of 'order with competition' in Japanese society (Pye 1985). Other cities in East Asia have not been as successful in combining order with competition. Excessive concentration, or the too rapid expansion of population and industrial activities, poses a serious threat to the quality of life in most cities and brings about a breakdown in urban management. The lesson from history is that order without competition tends to stifle innovation and diversification, whereas competition without order tends to bring agglomeration without centrality.

Forty years ago, Redfield and Singer asked whether modern colonial cities such as Hong Kong and Singapore could reverse their role from heterogenetic to orthogenetic centres. The answer may be yes. The leaders of Singapore, without a national hinterland, strive to create a great metropolis with an international hinterland in South-East Asia. Their growth triangle concept attempts to establish Singapore not only as an economic centre but also as a cultural centre (Koh 1989). However, Singapore's culture has been determined by the state, whose centrality is pervasive in national life. For all its advantages in producing order and stability for the people's social and economic welfare, the degree of predominance tends to discourage spontaneity and liveliness in people's thinking and action. Recognizing the drawbacks of a top-down social engineering approach, the Singaporean government encourages civic participation, hoping to create an organic community through a policy of 'liberalism from above' (Jones and Brown 1994). The government also supports traditional Asian values. The key issue is how to strike a balance between state intervention and a pluralistic society so as to encourage creativity and an imaginative approach to problem-solving.

Another important issue in the evolution of Singapore's national culture is whether the new urban middle class will subscribe to Confucian values that emphasize consensus or will give more value to pluralistic interests and individual rights.

Hong Kong presents a case of vibrant capitalism with minimum government intervention that has provided an economic haven not just for its inhabitants but for the overseas Chinese population as a whole. Detached politically from China and with a restricted border until the early 1980s, Hong Kong has established a separate urban identity, turning the cultural energy of its Chinese inhabitants into an economic force (Murphey 1984; Wong 1986).[7] A pattern of social stratification that differs from that of

---

[7] The history of Hong Kong was different from that of the Chinese treaty ports. The treaty ports were seen by the Western powers as beach-heads, agents for the transformation of China into a Western likeness through the power of commerce and industry. Ironically, the movement that successfully overthrew the old regime was founded in 1911 in the treaty ports. Here Chinese nationalism was born, primarily as a reaction to the humiliation of foreign imperialism. The Chinese Communist Party, founded in Shanghai in 1921, resisted foreign aggression and eventually expelled the foreign presence and all it stood for (Murphey 1984).

traditional China has also been created in Hong Kong. In China, social status was traditionally determined by qualification for office and upward mobility was based on a successful career as a government official, whereas in Hong Kong wealth by itself brings social status (Wong 1986). Freed from the rein of Chinese institutions, the people of Hong Kong have become entrepreneurs, concerned more with the welfare of their families than with that of the community. Without elaborate social engineering from above or explicit civic consciousness, Hong Kong has become a vibrant and thriving metropolis. What will be the effect when Hong Kong joins China in 1997? Perhaps the city will provide a model for capitalist development combined with a stress on family, one of the essential qualities of Confucianism.

The issue of achieving the correct balance of power between the state and the people applies to all East Asian cities where the government has been involved in economic and social development—how to sustain the spontaneity and vitality of urban residents without losing their social conscience and communal spirit. This is equivalent to the 'spontaneous self-diversification' that Jacobs (1961) emphasized in discussing the future of American cities.

The creativity of cites, as Eisenstadt and Shachar (1987) conclude, is derived from the heterogeneity of urban groups and their competition within the socio-political order. The emergence of new urban élites and subélites brings new cultural orientations, political concepts, values, and symbols, which in turn add vitality and diversity to the urban culture. In this context, the rise of the new urban middle class in East Asia is significant. Intellectuals, students, and other members of the middle class enjoy more autonomy than wage workers in the societies of Japan, South Korea, Taiwan, Hong Kong, and Singapore (Koo 1991; Jones and Brown 1994). Concentrated in large urban centres, they shape the dominant lifestyle and consumption patterns. They make new political and cultural demands, which often lead to debates and discussions because their needs and values are often ambiguous and contradictory (Cotton and van Leest 1992). They tend to hold traditional notions of hierarchy and social order that clash with their aspirations for an open, free society. The political ethos and way of life of East Asia's new urban middle class is still in a state of development.

The dynamic of competition among heterogeneous groups nourishes inventiveness and results in the development of civic consciousness and an urban culture specific to each particular city. The monolithic élitist structure of pre-modern cities of China and Korea, to some extent of Singapore, impeded the incorporation of new groups and thus new values and ideas. Extremes of social fragmentation have, however, led to the decline of morality and productivity in American cities, epitomized by the erection of walls around residential areas, workplaces, and public facilities (Harvey

1992). This is what de Bary (1994) termed 'coercive civility', a civility based on law backed by state power, as opposed to 'consensual civility', which is guided by reason, moral sense, and social discipline. Will East Asian cities develop and maintain a sense of consensual civility without being trapped in the inflexible social structures that discouraged diversity and spontaneity in the past or will they evolve into the divisive pattern of contemporary American cities, isolating one group from another, separating men from women, and tearing the family apart?

Conclusions

The industrialization imposed by the West and subsequent economic development in East Asia have unleashed forces of urban concentration sometimes beyond the control of governments and citizens. At the same time, the penetration of capitalist ideology and the increasing influence of 'world culture' are changing values, norms, and social relations. Cities are becoming increasingly dominated by international, commercial, and technological considerations, which are impossible to resist. The specific response of each city is a product of local and global interactions, including the city's historical and cultural character. Diverse responses, ranging from Beijing's quasi-market approach to Singapore's top-down management, indicate that a new urban identity is taking shape in these cities through the interactive processes of centrality and concentration.

One major difference between East Asian and Western societies is the Asian emphasis on loyalty to the collectivity such as family, company, and nation. Loyalty is highly valued in Asia, although the intensity and scope of loyalty varies between the countries. In East Asia, social relations are often based on trust and reciprocity with a long-term time horizon, whereas non-obligational contract is more prevalent in social exchanges in the West. Loyalty, when projected upward to the central authority, produces cultural cohesion. Commitment from below not only sustains the power of the central authority but also provides the energy to build the material base of that power.

Whereas loyalty and order are traits common to the Confucian tradition, they influence contemporary social relations and institutions differently in different East Asia cities. For example, the object of loyalty may vary with local circumstances and may shift over time.[8] In Hong Kong, a centre of

[8] Palais (1984) makes an interesting point concerning the potential contradiction of Confucianism. The emphasis on filial piety and loyalty could be used to subordinate the individual and the family to the ruler, but it might also function to encapsulate the family and protect it from the intrusion of state power (as evident in Hong Kong and to a lesser extent in Taiwan). Similarly, the belief in the primacy of the public good might lead to state control over land and wealth, but the emphasis on family solidarity might strengthen private property and reduce the state's power to control resources.

finance and management with limited centrality based on political order, loyalty is primarily to the family.[9] In Tokyo and Seoul, loyalty is directed to the workplace and to broader arenas of the political economy. An adaptive sense of community is one hallmark of Japanese society; whereas Korean society is characterized by a less pronounced sense of community mixed with an 'underground' ideology of individualism (Vogel 1987; Brandt 1987). In Beijing and Hanoi, *guanxi* (relations) operate in a diffused fashion, while the central state makes a claim for loyalty and self-sacrifice. In both the cities of East Asia and the societies at large, three key factors are essential for understanding the process of social change: (1) the balance between the state and the society; (2) the way the society incorporates heterogeneity (acceptance of heterogeneous groups and foreign ideas) and mediates conflicting interests; and (3) the way the society interprets and articulates the global impact on local culture.

This last aspect of social change will become more important as the global economy becomes more interdependent and competitive. Cities are more likely to compete with one another for their survival. An economic base, narrowly defined to include a city's resource endowments and industrial structure, is no longer adequate in this global era; rather, a city needs a cultural base to survive. This involves establishing or strengthening the centrality of the city, especially in managing social relations and the urban economy. The long-term competitiveness of a city, as for a nation, stems from institutional creativity rooted in the indigenous culture (North 1990; Schlack 1990; Thompson 1965).[10]

References

Beg, Muhammad A. J. (1986). Introduction, in Muhammad A. J. Beg (ed.), *Historic Cities of Asia*. Kuala Lumpur: Percetaken Ban Huat Seng.

Brandt, V. S. R. (1987). 'Korea', in G. C. Lodge and E. F. Vogel (eds.), *Ideology and National Competitiveness: An Analysis of Nine Countries*. Cambridge, Mass.: Harvard Business School Press.

Cavalcanti, Maria de Betania (1992). 'Totalitarian States and their Influence on City-Form: The Case of Bucharest', *Journal of Architectural and Planning Research*, 9: 272–86.

Chang, Ki-Ho, and Lee, Sok-Ho (1987). *Noja changja* (The philosophy of Laozi and Zhuangzi). Seoul: Samsung Publisher.

[9] We may define this family-centred loyalty as familist Confucianism, as opposed to statist Confucianism that places more emphasis on the concerns of the larger community. A parallel can be drawn with the contrast between Hinayana and Mahayana Buddhism.

[10] Fukuyama (1995) emphasizes the importance of social capital in economic growth, which is essentially a historical product of a society's culture.

Chang, Sen-Dou (1963). 'The Historical Trend of Chinese Urbanization', *Annals of the Association of American Geographers*, 55: 109–43.

Choe, Yong-Ho (1987). *The Civil Examinations and the Social Structure in Early Yi Dynasty Korea: 1392–1600*. Seoul: The Korea Research Center.

Chubu, Yoshiko (1967). *Kinseitoshino seiritsuto kouzou* (The formation and structure of modern cities). Tokyo: Shinseisho.

Cotton, James, and van Leest, Hyung-a (1992). 'Korea: Dilemmas for the Golf Republic', *Pacific Review*, 5: 360–9.

Cybriwsky, Roman (1991). *Tokyo: The Changing Profile of an Urban Giant*. London: Belhaven Press.

de Bary, William T. (1994). 'Multiculturalism and the Core Curriculum'. Cultural Studies Program, East–West Center, Honolulu.

Deuchler, Martina (1992). *The Confucian Transformation of Korea: A Study of Society and Ideology*. Cambridge, Mass.: Harvard University Press.

Dray-Novey, Alison (1993). 'Spatial Order and Police in Imperial Beijing', *Journal of Asian Studies*, 52: 885–922.

Eberhard, Wolfram (1956). 'Data on the Structure of the Chinese City in the Pre-industrial Period', *Economic Development and Cultural Change*, 5: 253–68.

Eisenstadt, S. N., and Shachar, A. (1987). *Society, Culture and Ubanization*. Newbury Park, Calif.: Sage Publications.

Fairbank, John K., and Reischauer, E. O. (1989). *China: Tradition and Transformation*. Rev. edn. Boston: Houghton Mifflin Co.

Fukuyama, Francis (1995). *Trust: The Social Virtues and the Creation of Prosperity*. New York: The Free Press.

Giddens, A. (1984). *The Constitution of Society: Outline of the Theory of Structuration*. Berkeley and Los Angeles: University of California Press.

Harvey, David (1992). 'Social Justice, Postmodernism and the City', *International Journal of Urban and Regional Research*, 16: 558–601.

Hashitani, Hirishi (1992). 'NIEs toshi Seoul no keisei' (The formation of a NIEs city, Seoul), *Chosenshi Kenkyukai Ronbunshu*, 30: 121–47.

Ho, P. T. (1959). 'Aspects of Social Mobility in China 1368–1911', *Comparative Studies in Society and History*, 1: 330–59.

Hoover, Edgar M., and Giarratani, Frank (1987). *An Introduction to Regional Economics*. 3rd edn. New York: Alfred A. Knopf.

Hou, Renzhi (1986). 'Evolution of the City Plan of Beijing', *Third World Planning Review*, 8 (5): 17.

Huntington, Samuel (1993). 'The Clash of Civilizations', *Foreign Affairs*, 72: 22–49.

Huus, Kari (1994). 'No Place like Home', *Far Eastern Economic Review*, 28 July.

Jacobs, Jane M. (1961). *The Death and Life of Great American Cities*. New York: Vintage.

——(1993). 'The City Unbound: Qualitative Approaches to the City', *Urban Studies*, 30: 827–48.

Jones, David M., and Brown, David (1994). 'Singapore and the Myth of the Liberalizing Middle Class', *Pacific Review*, 7: 79–87.

Kang, Man Gil (1973). *Chosonhoogi sangupjabonui baldal* (The development of industrial capital in the late Choson). Seoul: Korea University Press.

Kim, Song-Su (1985). *Ilcheha hankookkyongjesaron* (The economic history of Korea under the Japanese colonial rule). Seoul: Kyongjinsa.

Kim, Yong-Ock (1991). *Noja cholhak igoshida* (This is Laozi philosophy). Seoul: Tongnamu.

King, Anthony D. (1984). *The Bungalow*. London: Routledge & Kegan Paul.

——(1989). 'Colonialism, Urbanism, and the Capitalist World Economy', *International Journal of Urban and Regional Research*, 13: 1–18.

Ko, Sung-Je (1988). *Hankook sahoekyongjesaron* (The socio-economic history of Korea). Seoul: Ilchisa.

Koh, Tai Ann (1989). 'Culture and the Arts', in K. Singh and P. Wheatley (eds.), *Management of Success: The Moulding of Modern Singapore*. Singapore: Institute of Southeast Asia.

Koo, Hagen (1991). 'Middle Classes, Democratization, and Class Formation: The Case of South Korea', *Theory and Society*, 20: 485–509.

Lau, D. C. (1981). *Lao tzu Tao te ching* (The moral teachings of Laozi). Harmondsworth: Penguin Books.

Lawrence, Denise L., and Low, Setha M. (1990). 'The Built Environment and Spatial Form', *Annual Review of Anthropology*, 19: 453–505.

Lee, Ki Back (1988). *Hankooksa shinron* (The new Korean history). Seoul: Ilchisa.

Meyer, Jeffrey F. (1991). *The Dragons of Tiananmen*. Columbia: University of South Carolina Press.

Mote, F. W. (1977). 'The Transformation of Nanking, 1350–1400', in W. G. Skinner (ed.), *The City in Late Imperial China*. Stanford, Calif.: Stanford University Press.

Mumford, Lewis (1961). *The City in History*. London: Secker & Warburg.

Murphey, Rhoads (1984). 'City as a Mirror of Society: China Tradition and Transformation', in J. Agnew *et al.*, *The City in Cultural Context*. Boston: Allen & Unwin.

Nhuan, Nguyen Duc (1984). 'Do the Urban and Regional Management Policies of Socialist Vietnam Reflect the Patterns of the Ancient Mandarin Bureaucracy?', *International Journal of Urban and Regional Research*, 8: 73–89.

North, D. C. (1990). *Institutions, Institutional Change, and Economic Performance*. Cambridge: Cambridge University Press.

Palais, J. B. (1984). 'Confucianism and the Aristocratic/Bureaucratic Balance in Korea', *Harvard Journal of Asiatic Studies*, 44: 427–68.

Pye, Lucian W. (1985). *Asian Power and Politics*. Cambridge, Mass.: The Belknap Press of Harvard University Press.

——(1990). 'China: Erratic State, Frustrated Society', *Foreign Affairs*, 69: 56–74.

Redfield, R., and Singer, M. (1954). 'The Cultural Role of Cities', *Economic Development and Cultural Change*, 3: 53–73.

Rozman, Gilbert (1978). 'Urban Networks and Historical Stages', *Journal of Interdisciplinary History*, 9: 65–91.

Samuels, Marwyn S., and Samuels, C. M. (1989). 'Beijing and the Power of Place in Modern China', in John Agnew and James Duncan (eds.), *The Power of Place*. Boston: Unwin.

Sato, Seizaburo (1979). 'Response to the West: The Korean and Japanese Patterns', in Albert M. Craig (ed.), *Japan: A Comprehensive View*. Princeton: Princeton University Press.

Sawyer, Larry (1975). 'Urban Form and the Mode of Production', *Review of Radical Political Economics*, 7: 52–8.

Schlack, Robert F. (1990). 'Urban Economies and Economic Heterodoxy', *Journal of Economic Issues*, 24: 17–47.

Skinner, W. G. (1977). 'Introduction: Urban Development in Imperial China', in W. G. Skinner (ed.), *The City in Late Imperial China*. Stanford, Calif.: Stanford University Press.

Smith, Henry (1978). 'Tokyo as an Idea: An Exploration of Japanese Urban Thought until 1945', *Journal of Japanese Studies*, 4: 45–80.

——(1979). 'Tokyo and London', in Albert M. Craig (ed.), *Japan: A Comparative View*. Princeton: Princeton University Press.

Smith, Thomas C. (1966). 'Japan's Aristocratic Revolution', in Reinhard Bendix and Seymour Martin Lipset (eds.), *Class, Status, and Power*. New York: Free Press.

Sohn, Jung Mok (1977). *Choson Sidae Toshisahoe Yongu* (The study of urban society in the Choson period). Seoul: Ilchisa.

Spodek, Howard (1993). 'Beyond Rorschach Tests: Palimpsests and Nodes, Conflicts and Consciousnesses in South Asian Urban Theory', in Howard Spodek and Doris M. Srinivasan (eds.), *Urban Form and Meaning in South Asia: The Shaping of Cities from Prehistoric to Precolonial Times*. Washington: National Gallery of Art.

Strand, David (1989). *Rickshaw Beijing: City People and Politics in the 1920s*. Berkeley and Los Angeles: University of California Press.

Thompson, Wilber R. (1965). *A Preface to Urban Economics*. Baltimore: Johns Hopkins University Press.

Tilley, C. (1984). 'History: Notes on Urban Images of Historians', in L. Rodwin and R. M. Hollister (eds.), *Cities of the Mind*. New York: Plenum Press.

Toynbee, Arnold (1967). Introduction, in A. Toynbee (ed.), *Cities of Destiny*. New York: McGraw-Hill.

Tu, Wei-Ming (1992). 'Intellectual Effervescence in China', *Daedalus Journal of the American Academy of Arts and Sciences*, 121 (125): 9.

Vogel, E. F. (1987). 'Japan: Adaptive Communitarianism', in G. C. Lodge and E. F. Vogel (eds.), *Ideology and National Competitiveness: An Analysis of Nine Countries*. Cambridge, Mass.: Harvard Business School Press.

Walton, John (1984). 'Culture and Economy in the Shaping of Urban Life: General Issues and Latin American Examples', in J. Agnew, J. Mercer, and D. Sopher (eds.), *The City in Cultural Context*. Boston: Allen & Unwin.

Wong, Siu-Lun (1986). 'Modernization and Chinese Culture in Hong Kong', *China Quarterly*, 106: 306–25.

Yoshida, Mitsuo (1992). 'Kanjono toshiokukan' (Urban space in old Seoul), *Chosenshi Kenkyukai Ronbunshu*, 30: 91–120.

APPENDIX

List of Major Dynasties and Kingdoms in East Asia

China	Japan	Korea	Vietnam
Yuan 1271–1367	**Kamakura** 1185–1335	**Koryo** 918–1391	**Tran** 1226–1400
Ming 1368–1662	**Ashikaga** 1336–1598	**Yi** 1392–1909	**Later Le** 1428–1777
Ching 1663–1911	**Tokugawa** 1600–1867		**Nguyen** 1800–1883
Kuomintang 1928–1948	**Meiji** 1868–1912	**Japanese rule** 1910–1945	**French rule** 1884–1945
People's Republic and Taiwan 1949–	**Taisho** 1912–1926 **Showa** 1926–1988	**North and South Korea** 1945–	**North and South Vietnam** 1945–1974

Source: Adapted from Fairbank and Reischauer (1989).

3

Urbanization and Social Transformations in East Asia

Mike Douglass

'Change Life!' 'Change Society!'
   These precepts mean nothing without the production
of an appropriate space.
                                                    (Lefebvre 1991: 59)

The concern of the discussion that follows is the relationship between culture, social change, and power and the city in the context of an increasingly globalized economy. The aim is to clarify how contemporary issues affect nation-states and territories in East Asia, each of which in its own way is facing a turning-point in development.[1] The general theoretical framework that is developed combines a world systems[2] perspective with insights from recent writing on 'structuration'[3] and 'cultural-context'[4] on the urban condition in East Asia. Contemporary urbanization processes and their human consequences in East Asia are seen as being outcomes of interaction between each social formation and the global capitalist system.

   On the global scale the formation and transformation of the world economy particularly since the height of European imperialism in the nineteenth century has seen the penetration of a profoundly new logic of social, political, and economic organization into East Asian societies. Among the key

[1] Unless otherwise stated, East Asia refers to Japan, Korea, China, Taipei, Hong Kong, and, due to ethnic and cultural influences, Singapore.

[2] 'World system' as used here is not based on Wallerstein's (1979) well-known world system thesis and its heavy reliance on core, periphery, and semi-periphery categorization of nation-states. Rather, it is intended, first, to counter much of received urban theory, which asserts that urbanization can be understood as a 'national' rather than international process, and, second, to include epochal changes in the global economy as a principal dimension (but not determinate) of localized process of change and development, including urbanization (King 1989; Corbridge 1986; Douglass 1993*a*).

[3] Following from Giddens's theory of structuration (1984), space is an integral dimension of social behaviour and is largely played out at very localized scales where the action of the individual is most directly linked to socio-cultural structures and larger systemic forces.

[4] The 'cultural-context' approach seeks an explanation of urban form, growth, and life through cultural institutions (Agnew and Duncan 1989: 277).

outcomes of this penetration are the formation of the modern nation-state based on division of state apparatus into central bureaux rather than hierarchies of territorial domains, the development and expansion of market forces that have inverted relations between society and economy and have become increasingly extroverted through international linkages, a social division based on economic class rather than ascribed status, and urban forms oriented toward servicing national–international economic linkages rather than sacerdotal, juridical, military, or administrative principles of spatial organization.

A focus on the international scale also shows how the world system itself changes through time. Although both mainstream neo-classical and Marxist theory continue to assume an unvarying (world) capitalist system, the contemporary international economy is fundamentally different from that which prevailed even two decades ago during the Fordist era (Corbridge 1986; Dunford 1990).[5] The critical differences include the explosive transnationalization of capital now operating through international circuits of finance, production, and commerce, and, second, the parallel development of a new international division of labour, decentralizing segments of the global assembly line to selected Third World sites. Both have placed new pressures on every society to restructure its economy, dismantle or put off constructing state welfare systems, accept new forms of flexible production and relations between labour and capital, and otherwise create new approaches toward attracting footloose global capital. The newly industrialized economies of East Asia (Asian NIEs) are no exception. Although apparent principal beneficiaries of the global decentralization of labour-intensive production in the 1960s and 1970s (and more recently in China), domestic firms in these economies are also moving offshore, creating similar pressures for economic as well as political restructuring.

The impingement of powerful external forces upon the capacity for autonomous changes within a given society does not, however, yield uniform outcomes across all societies, but is instead filtered through historically differentiated constellations of socio-cultural, economic, and political struc-

---

[5] What has been called the Fordist period can be understood as being composed of (1) a system of manufacturing based on the fragmentation of the labour process into simple assembly-line operations that (2) was paralleled by unionization of labour and the adoption of Keynesian economic management and welfare guarantees by the state to allow for the emergence of mass consumer markets (Esser and Hirsch 1989). The principal achievement of this period, which was responsible for post-Second World War world economic expansion up to the late 1960s, was its capacity to combine increasing labour productivity with increasing wages and thereby forestall a consumption crisis that would have resulted from the quantum increases in manufactured goods without commensurate increase in effective demand. The advent of the current post-Fordist era followed from the collapse of the Fordist systems of production and regulation that followed from the movement of the lower end of assembly-line production to the Third World, the reorganization of higher-order manufacturing into systems of flexible production based on interlinked, small scale firms, the decline of organized labour, the deregulation of industry, and the dismantling of welfare support programmes beginning in the 1970s.

tures that, in turn, produce significant variations in (implicit) development models. These, too, experience fundamental shifts at critical junctures. As summarized by Corbridge, 'Accounts of differential development must recognise that the dynamics of a changing capitalist world economy are always mediated by conditions of existence . . . which vary in space and time and which are not directly at the beck and call of a grand "world system"' (1986: 245–7).

The city experiences parallel shifts, and this view of urbanization as part of a process of transformation at both local and global scales stands in contrast to three other well-known perspectives on economic development and social change. First, it argues against the modernization school of thought that has developed within Western sociology and neo-classical economics, which proposes that all nations follow a single, linear development path from 'traditional' to 'modern', with urbanization as a principal force in realizing increasing economic 'rationality' in decision-making, efficiency through specialization on the basis of comparative advantage for international trade. It rejects the notion that every society can, will, or should follow a *laissez-faire* development path attributed to the advanced Western nations. In fact, this was not the experience in the West, where the *laissez-faire* period of market competition that appeared briefly in the nineteenth century rapidly gave way to monopoly capital and oligopolistic competition and where the Keynesian state guided economy and society through the post-Second World War Fordist period of industrialization. Nor has it been the case in East Asia, where the state has been exceptionally interventionist in stimulating economic growth (Amsden 1989; Douglass 1994). It also finds little historical validity in the thesis allied with modernization theory that once an economy achieves 'take-off' its trajectory toward industrialization and beyond is secure.

Secondly, the perspective put forth here stands in contrast to the Marxist structuralist and world systems views that implicitly, if not explicitly, assert that local development is wholly determined by global capitalist development. The position taken rests on the premise that each society enjoys a degree of relative autonomy from these forces. Borrowing from Marxist vocabulary, this position is oriented toward the study of specific social formations rather than an abstract modelling of the articulation of a generalized pre-capitalist with a dominating world capitalist mode of production. A focus on differences in localized histories (i.e. engaging in 'locality' studies) reveals the resilience and potency of socio-cultural institutions, the national and local state, and other institutional arrangements within each society in filtering external imperatives, managing economic and other crises, and otherwise creating varied modes and paths of development.

Finally, the themes elaborated below also counter the rhetoric currently being popularized in several Asian countries that asserts the existence of an overarching 'Asian' way of development that is both commonly shared among all Asian societies and fundamentally distinct from an assumed Western approach. In a similar vein, the idea of a 'neo-Confucianist' model of development unique to a subset of Asian countries and territories is not compelling. While culture histories have been shared among these societies, their long periods of relative isolation from each other, followed by the profound impacts of colonialism and, more recently, the contrasting ways in which each has been reinserted into the world economy as exporters of manufactured goods, have resulted in substantial differences in cultural norms and practices as well as in economic institutions, the role of the state, patterns of social differentiation, and, therefore, the urban social fabric and built environment. The assertion that there exists an indigenous pan-Asian paradigm of urbanization can be maintained at best only at the highest and most superficial levels of abstraction and with great risk of substituting ideological positioning for historical analysis.

In sum, the examination of the interplay between socio-cultural institutions and urbanization is best understood within a global–local framework that accepts a more open-ended, historically contingent process of development than is allowed by received modernist or Marxist theories or 'Look East' development ideologies. Within this framework, five themes serve to organize the discussion that follows. They are:

1. Urban form and function are both socially constructed and socially organizing. Cities promote as well as constrain processes of socio-cultural, political, and economic change and development.
2. Cultural influences among East Asian societies have not resulted in a single East Asian or a neo-Confucianist realm of beliefs, ideals, or practices but have instead led to striking socio-cultural difference.
3. The transformations accompanying urbanization are historically episodic and have been the outcome of internal contradictions and crises, the penetration of external forces and/or fundamental changes in the world capitalist system, or a combination of both.
4. Cities contain multiple landscapes. No single interpretation or vision of the city and its constituent elements is totalizing. This presents profound dilemmas in the politics of planning for the future of cities in East Asia that are inseparable from political change and reform toward the inclusion of many voices.
5. Societal guidance of development in East Asia will increasingly focus on restructuring the city, both as the built environment for human endeavours and as the site for the formation of a more inclusive political process.

Together these themes suggest the possibility of much more room for alternative paths of development than received theory has allowed, but they also suggest that constructing such alternatives confronts formidable local as well as global obstacles.

Socio-cultural Change and the City in History

The city is not simply a backdrop in the theatre of social change, but, through both functional support of human activities and its representations of symbolic and socio-cultural meaning, it is an integral part of system maintenance and social change alike. Cultural and social continuity, particularly with regard to the formation and endurance of social hierarchies, is contingent upon 'the social reproduction of rules and power relations, and upon the parallel emergence of a socially produced spatial structure' (Pred 1984: 254).[6] Or, as explained by King (1984, 1989), the city is simultaneously a product and a producer of social organization. The built environment of the city serves to define the use of space, creates and organizes work, and supports new economic, social, political, and cultural practices just as much as it is an outcome of the interplay of social forces. The city, in the terminology of Castells (1977), is not just a reflection of society, but is one of its material dimensions.

The interplay between the city and social change is revealed in even the earliest periods of urbanization; for the origin of cities is the origin of complex societies capable of spatially organizing the flow of resources and wealth over geographically extensive regions and empires from densely settled urban centres (Abu-Lughod 1991). In addition to the direct use of military force, the sources of power organizing the expansion of these cities and their hinterlands were found in cultural institutions that legitimized both the position of urban élites and the superiority of urban centres over hinterlands. The creation of cities was thus required not only for the purposes of physically sustaining increasingly complex societies and creating key nodes from which to organize surplus appropriation from distance places, but also symbolically to reify the position and status of dominant socio-cultural institutions and the power they conveyed to select groups and segments of society. Religious, cosmic, juridical, and other principles for architectural and urban design forms were adopted to represent the hegemony of socio-cultural institutions and provide emblems and icons for the hierarchical structure of social relations (Carter 1983; Johnson 1970).

Mesopotamia in 3500 BC exemplified this early historical process of heightened social differentiation with urban expansion. With the city centred on the temple, the priesthood—through its assumption of the role of

---

[6] In other terms, the construction of urban habitats 'always involves an appropriation and transformation of space and nature that is inseparable from the reproduction and transformation of society in time and space' (Pred 1986: 337).

intermediary between mortals and the deities—legitimized its control of communal land and its rights to command resources from beyond its own surrounding hinterland. Ancient cities in Egypt, China, South Asia, and Central America were based on similar socio-cultural structures that used imperial cities to collect tribute from more distant centres and territories. Cultural institutions and the power structures they supported became urban centred in the process of extending territorial domains and, in part, were used to direct the building of the city, including land development, the provision of economic and social infrastructure, construction of residential quarters, and securing transportation and communications linkages with the 'outside' world. Monumental temples with storerooms for accumulated wealth, elegant palaces for deity-rulers, walled quarters, and gated entrances to the urban settlement were among the ensembles of structures in the cities planned under the mandate of culturally and spiritually invested leaders and seers. Whether priests, military commanders, juridical leaders, or other keepers of the realm, these élites applied social philosophies and cosmic explanations alike to build and imbue cities with cultural meaning.

Urban form took on representations of these complex societies in which cultural mores and institutions were also called upon to define and enable the division of the city into designated areas in which various groups and classes would live. The occupation of the centre of the city, to which access was forbidden or severely limited to all but the most privileged élites, represented the highest citadels of culture and was used to signify in location and architecture the concentration of power among the élite. Non-agricultural pursuits grew along with an increasing complexity in the social division of labour, with many specialities no longer being essential for the direct material functioning and reproduction of society but of great importance to sustaining the continuity of cultural values and social structure.

In this interplay culture matters greatly in societal processes, including urbanization. Contrary to economistic treatments of urbanization that have tended to dominate human geography, and despite the trivialization of culture in much of urban sociology caught up in theories of modernization, culture is a key dimension of social change as well as urban development in all societies. More than being a repository for commonly held beliefs, cultural norms and culturally defined ways of acting are equally a source of social organization and mobilization that will continue to differentiate the urban condition in East Asia even as these societies become both highly urban and increasingly integrated into global socio-political and economic spheres of power.

Defined as expected ways of behaviour that are encoded in social, political, and economic institutions, culture forms 'networks of shared practices' (Agnew and Duncan 1989: 1) and 'systems of symbols' (Rapoport 1984: 50)

that serve to maintain social structures, including status, social differentiation, and access to wealth and power. They also invariably contextuate even
radical social transformations. Urban form, design, and architecture play a
significant role in symbolically representing socio-cultural institutions and,
as stated, thereby give the city the purpose of not only functioning as an
efficient means for spatially organizing urban and regional life, but of also
contributing to the ideological underpinnings of dominant forms of social
organization and associated structures of power.[7]

The history of Beijing illustrates this theme. From 1270 to 1912 Beijing
was a highly regimented celestial city with a 'seeming endless maze of
walled compounds within walled compounds within walled compounds that
were imbued with the signs of power, authority and hierarchy' (Samuels
and Samuels 1989: 212).[8] It was at the same time a functioning city of
merchants, traders, sojourners, and lower segments of society performing
the necessary tasks of reproducing the city's existence (Murphey 1984).[9]
Rather than fragmenting the symbolic content of the city along social
divisions, the increasing scale and diversity of the city and its population
may have been actually led to greater efforts to invest more energy and
resources to maintain the hegemony of the dominant socio-cultural and
political institutions through the increasing emphasis on symbolic content.
This would follow from observations that, first, social inequality and divisions lead to higher investments by élites in the symbolic content of the built
environment: 'The structural investment in symbolic functions increases in
response to greater social differentiation. In societies having more groups
and more social distinctions, there is a need to communicate even more
information materially. Furthermore, as social units become increasingly
specialised, artefacts with high symbolic content are needed to integrate a
society's disparate parts' (McGuire and Schiffer 1983: 281).

This was surely the case with Beijing and other great pre-colonial cities in
Asia. As they became more complex, the grandeur of symbolic architecture
and urban form that focused on the Confucianist élite also excelled. But
by the mid-nineteenth century the 'whole fabric' of Confucianist China
began to unravel, first from Western colonial penetration and revolutionary
nationalism, which ended the Qing dynasty in 1911, and again in the

[7] As summarized by Walton (1984: 79), '[culture] is the set of beliefs that are acted out in the
political and economic struggle in a process of reciprocal influence'.
[8] Although McGuire and Schiffer (1983) argue that symbolic functions representing cultural
beliefs and institutions do not contravene utilitarian functions, in the case of Beijing, the
division of the celestial city in a manner that prevented direct access by all but the highest
levels of the imperial household to cross from one side of the city to the other must have been
a significant interference in the functioning of the city's economy.
[9] Thus the 'traditional' cities appearing in the Ch'in and Han dynasties were functionally
'designed to control and tax the countryside' and 'the primary responsibility of officials was to
ensure the productivity as well as the orderliness of the agricultural countryside, since it was
this which sustained the empire, its power, its cultural grandeur, and its bureaucracy' (Murphey 1984: 190).

mid-twentieth century with the capture of the state by the Communist Party. With the advent of each of these epochs, Beijing experienced 'wholesale transformation', and by the mid-twentieth century the Confucianist imprint no longer dominated the expansion of the city.[10]

Yet throughout this century attempts have been made to reassert inherited cultural symbols into the built environment of cities, as exemplified by the Guomindang Party's 'New Life Movement' of the 1930s calling upon urban archaeology to rediscover China's 'design authenticity' and thereby to legitimize its position against the iconoclasm of the Communist Party (Samuels and Samuels 1989). The more striking paradox was Mao's urging for China to 'master its own style' through incorporation of traditional architecture into socialist society. And, as China turns toward a more market-oriented economy, renewed efforts are now being given to preserving courtyard housing, walled neighbourhoods, and other urban forms symbolizing the continuity of culture in the city.[11]

The understanding that the culture–city relationship is a critical feature of historical change in East Asia points toward the future as well. Following from the earlier work of Castells (1977), as these societies become highly urbanized and labour organized around the factory system declines, collective consumption in the form of the built environment of the city is increasingly becoming the focus for state–civil society contention as well as class conflict in East Asia. In this shift, finding the 'moral high ground' calls upon cultural values and often focuses on specific sites in the city to symbolize the historical mission being undertaken (Apter and Sawa 1984). The playing out of these contentions over the built environment is not directed at the city alone, but is equally the political dimension of social change that is reconstituting the basis for future development of these societies.

Cultural Filtering and Layering

The second theme takes as a point of departure the argument put forth by some scholars that contemporary development (and by extension urbanization) in all East Asian societies can be distinguished from that of other regions of the world under a unifying concept of 'neo-Confucianism'

[10] By the early 1930s there were hospitals, office buildings, hotels, schools, theatres, shops, department stores, workshops, factories, public parks, sewage systems, trolley lines, electric and telephone lines—the 'modern city' emerged to dominate the celestial city (Samuels and Samuels 1989).

[11] Such aims to use urban form to preserve a culture already distanced from its pre-colonial roots and to resist emergent market forces are, however, increasingly difficult to realize as the institutionalization of land markets results in the conversion of central areas into high-rise retail and commercial building sites in China's cities. Also, in an era of 'Let some get rich first' government slogans, the social mission of these efforts is also suspect. Rather than providing a culturally sensitive urban habitat for 'the people', the more likely outcome is to create gentrified neighbourhoods for élites and *nouveaux riches*.

(O'Malley 1988; Hsiao 1990). Vogel (1991) goes even further by asserting that the Asian NIEs all share an 'industrial neo-Confucianism' that underlies their accelerated export-oriented manufacturing growth. The main features of this proposed cultural commonalty include meritocracy within the state élites, recruitment through education, strong vertical and group loyalty, and emphasis on personal improvement throughout life. Labelled 'Post-Confucian' by Moody (1988), the cultural traits shared by the Asian NIEs (and presumably Japan and China as well) present a code of ethics to guide behaviour that is institutionalized, in part, by a centralized bureaucracy which is essentially authoritarian in nature and is allowed to formulate policy goals independently of particular groups.

While Confucianist ideology and socio-cultural practices are readily apparent in East Asia, the degree of similarity among them is overstated and misses the very real differences among these societies. Neo-Confucianism itself is not as uniform as some authors imply.[12] Moreover, ancient as well as more contemporary influences and overlays of culture have created much more complex societies that cannot easily be reduced to a single cultural type.

What is at stake in the debate on the prevalence of an overriding set of cultural norms is a variation of the question of whether locally constituted social formations (i.e. nation-states or colonial states) matter in any significant way in the course of history. In the case of East Asia, the centuries of extreme isolation of both Korea and Japan from active interaction with China makes this debate all the more striking. To suggest that influences from China prevailed without significant transformation during the 600 years of the Hermit Kingdom of Korea or the nearly 300 years of the closed-door policy of the Tokugawa shogunate is to take a highly simplistic view of social change. Cultural diffusion is more accurately a process of 'filtering' through socio-cultural, economic, and political relations and a layering of newly innovated forms over historically developed ones.

This filtering is manifested in the building of cities. Thus although Nara and Kyoto were planned as copies of the capital of the Tang dynasty of China, key differences in social structures eventuated in marked differences in urban form. The relative weakness of the Japanese emperor left the Imperial Palace in design and in practice to be more of a cloister for living deities than the epicentre of political power. Similarly, in Korea the ascendance of neo-Confucianism with the founding of the Yi dynasty was not simply the borrowing of traditions from China, but was much more intensively focused on solidifying the dominance of the *yangban* class and restoring a centralized power that had been dissipated under the previous Koryo dynasty. Unlike China, for example, Buddhism was largely purged from

---

[12] Morishima (1982), for example, details how the Confucian concept of loyalty varies greatly among Japan, Korea, and China.

urban life in Korea, resulting in cities devoid of Buddhist temples (Han 1974: 210).

To restate the case, while urbanization and social change are subject to a variety of forces operating above the level of a given social formation or society, history is fundamentally contingent upon the playing out of histories in real geographical space. In the same vein, the future of cities and societies in East Asia will also be contingent upon the ways in which global processes are localized.

Social Change and Urbanization as Episodic Transformations

The third theme is that the societal transformations accompanying urbanization are episodic or epochal in nature. Whether produced by internal contradictions in societies that reach points of crisis that mobilize anti-systemic forces to previously unexpected levels of power, as in the case of the Meiji Restoration in Japan, or induced by the penetration of powerful external forces, as in the creation of Hong Kong and Singapore as entrepôts for European imperialism, or through a conjuncture of both, the significance of these historical processes is that social change cannot be theorized as being either a product of incremental change along a linear development path or an outcome of progress through predetermined stages from a lower to a higher level of development.

Among the most important episodes affecting urbanization in Asia was the appearance of Western, and later Japanese, imperialism. Contemporary urbanization in East Asia cannot be understood without a thorough grasp of the impact of this epoch on national urban patterns, urban form, and the new urbanized classes of people and social relations. European imperialism, which lasted from the sixteenth to the mid-twentieth century and reached its highest plateau in the late nineteenth century, was almost singularly responsible for creating from coastal villages and small settlements what are now the largest cities in Asia. Rangoon, Jakarta, Bandung, Saigon, Manila, Hong Kong, and Singapore were all colonial creations, and other cities such as Shanghai, Seoul, and Taipei were heavily influenced by the roles assigned to them in the colonial economy. New ethnic mixes of people induced by colonially supported international labour migration, new classes of comprador capitalists and wage labour, the paradoxical empowerment of local officialdom to effect indirect rule, and the stimulus for international trade were all part of the colonial experience that, in the terminology of Redfield and Singer (1954), created 'heterogenetic' cities that stood in opposition to the traditional 'orthogenetic' ones.

Given the profound impact of colonialism, which not only restructured territorial patterns of urbanization but also subordinated indigenous cultures to the logic of world mercantilism, it is all the more surprising how this period has been under-studied in terms of urban form (King 1989). The colonial period had at least five major impacts: transformation—but not elimination—of indigenous socio-cultural institutions to serve the interests of colonial economies; creation of centres for accumulation on a global scale that necessitated the physical restructuring of cities; the establishment of new official forms of urban planning; a heightening of uneven social and spatial development in host territories; and opening of cities to new peoples as well as creating new urban classes.

The first change, namely, dominance and transformation without dissolution of pre-colonial culture and social relations, is one of the most important dimensions of urbanization of that period (Rey 1971). In many instances, puppet governments and the incorporation of pre-existing élites and political rules into colonial systems actually increased the power of indigenous élites and sanctioned the construction and use of the built environment to house and symbolize pre-colonial cultural relations. Palaces, rather than being destroyed, were given even greater grandeur as the cultural integrity once supporting them from within was weakened, 'mutated', and subordinated to external forces. Elsewhere, 'native quarters' were established on the periphery of cities while foreign élites occupied central areas (King 1989). Recruitment of local people into various colonial activities, ranging from merchant businesses to coolie labour, also relied upon local cultural institutions (Basu 1985; Deyo 1990).

The reliance on indigenous socio-cultural institutions and urban form was nevertheless partial. The greater project at hand was to create centres for rural resource exploitation and appropriation of surpluses for colonial powers. The city was mobilized and remade for this purpose. As summarized by McGee (1967: 56–7) in his seminal work on colonial cities in South-East Asia: 'The most prominent function of these cities was economic; the colonial city was the "nerve centre" of colonial exploitation. Concentrated there were the institutions through which capitalism extended its control over the colonial economy—banks, agency houses, trading companies, shipping companies.'

New forms of architecture, such as the 'bungalow', appeared (King 1984); statues of European nobility and icons of European power were erected and placed in front of colonial buildings, parks, and other 'public' spaces. This was all partly made possible by the third, and equally remarkable, impact of colonial rule, namely, the institutionalization of bureaucratic systems of governance, including city planning and management. By the end of the colonial period following the Second World War, many colonial cities had established substantial bureaucracies that were responsible for such urban functions as sanitation, roads and transportation, and commerce. They also reached deeply into the countryside to rule over or replace localized systems of governance with a new district officialdom. Functional lines of authority extended outward to replace territorial ones, allowing rule from the centre to be more effective than it had been in many of the pre-colonial regimes.

Fourth, colonial rule also tended to heighten spatial imbalances of power and consumption. The development of coastal cities often shifted the locus of power from centres located in rural heartlands, and, through substantially increased levels of rural exploitation, created new primate cities that also received the bulk of investment in modern infrastructure and services. Regional and rural–urban gaps intensified as primate cities emerged to cater both to world markets and to new groups and classes in them.

Finally, all of the above brought about a substantial social restructuring of cities. Among the most profound impacts of colonialism was the filling in of new urban occupations by non-indigenous people. In South-East Asia, the reluctance of indigenous populations to sell their labour for a daily wage led to the wholesale migration to South-East Asian cities of Chinese who brought with them not only their labour power, but also business acumen that was manifested in the appearance of easily recognized Chinese residential and business sectors of the city that soon dominated petty commodity production and provided key business functions sustaining the colonial economy. In other countries, namely China, quarters were developed for Europeans. In Korea, Japanese colonialists not only took over official functions, but also usurped land from Korean landlords and peasants to become farmers themselves. Wherever colonialism took hold, new cultural and ethnic mixes were induced and new classes of people—waged labour, petty capitalists, rudimentary urban middle class—appeared to fill the needs of the colonial economy.

The post-colonial period of urbanization and development in East Asia has been even more profound. Pre-colonial and colonial processes and patterns of development tend to pale when compared to post-colonial urbanization in Asia. In most instances, cities accounted for less than 20 per cent of the national population in East Asia in 1950. By the year 2000, most countries will have half or more of their populations living in cities. Several will have already completed their urban transition by reaching levels of 70–80 per cent urban by that year. These trends are marking the latter half of this century and the beginning decades of the next century as the most compressed and intensive period of urbanization in Asian—and world—history. With very high national population growth rates during much of this period, and with continued polarization of development in a few urban regions, dominant metropolitan regions continue to double in size every fifteen to twenty years and are already reaching 'science fiction' population sizes of more than 20 million.

Such quantum increases are resulting in the emergence of extended urban regions that dwarf their past in terms of physical scale (Ginsburg *et al.* 1991). They are also part of an epochal transformation in socio-cultural institutions and relations. Since the late 1960s the articulation of past with the present has become imbedded in national efforts of 'late industrialization' and integration into the world capitalist system that is marked by its own spatial logic (Douglass 1993*b*, 1994). The rapid transnationalization of capital, the emergence of a new international division of labour integrating Third World labour into the global factory, and the incentives for foreign investment and other hospitality constructed in Third World societies have all led to new opportunities—and new imperatives—for accelerated economic growth based on export-oriented manufacturing that many Asian governments have actively pursued over the past quarter of a century (Douglass 1991, 1992).

Just as the temple or palace occupying the centre of the pre-colonial city gave way to merchant capital and colonial administrative centres from the seventeenth to the early twentieth century, colonial cities with their now quaint buildings are yielding to the modern commercial skyscraper, high-rise apartments, massive automobile-supporting road systems, and mega-infrastructure such as international airports and deep-draught harbours as the means to join networks of 'informational cities' participating in and commanding attention in the global circulation of capital.

Social relations changed in parallel with changes in the built environment. Peasants have been brought into the city to become the new industrial proletariat. Particularly where export-oriented manufacturing has 'taken off', rural women are recruited wholesale to fill low-wage assembly-line work in light industry. In higher-income economies such as Japan, Korea, and Taiwan, the spectre of foreign labour filling such jobs has appeared in recent years. In other settings, such as Hong Kong, Filipinas have been brought in by the tens of thousands to fill domestic servant occupations.

With regard to theories of social and political change, the rapid expansion of what was a relatively small urban middle class is one of the most often-cited features of the first post-Second World War era of accelerated urban-industrial growth in East Asia. The importance of this phenomenon is great. The appearance and political mobilization of this class has been credited with the emergence of democratic movements and new demands for widening political community to include voices from civil society outside the state. Some have even proclaimed this process to be the signal event in the discovery and legitimization of civil society in these societies (Cotton 1992).

Secondly, the emergence of a significantly large middle class is also linked to the transformation of cities from 'theatres of (global) accumulation' (Armstrong and McGee 1985) to centres of consumption moving toward a par with cities in the West and Japan. In parallel, the form and architecture of the city are seen as increasingly representing a convergence of consumption on a world scale. This is attested to by the appearance in almost all market-oriented cities in East and South-East Asia of in-town and suburban shopping malls, branches of international supermarket chains, condominiums designed with all modern 'Western' features, and boutiques specializing in *haute couture* designed in Italy and France, but with some irony fabricated by erstwhile peasants in garment factories in Asia.

The result of all of these contemporary changes is, in its most benign form, an overlay of old by new cultural and economic landscapes in the city. But in many cases the results have been more extreme, with modern office towers and urban industry not just sitting beside older temples and palaces, but actually replacing them. The hyper-growth of cities and their dependence on industrial labour has also resulted in land and housing crises of unprecedented proportions. While some governments have remained unresponsive to this crisis, others, such as those of Singapore, Hong Kong, and Korea, have created new landscapes of mass housing that now compete with high-rise commercial buildings to dominate the visual image of the city.

Despite the seemingly overwhelming transformations induced through a common strategy of export-oriented industrialization, the impingement of powerful external forces upon the capacity for autonomous changes does not yield uniform outcomes across all societies. As argued earlier, these forces are instead filtered through locally constituted and articulated constellations of socio-cultural, economic, and political structures that, in turn, produce significant variations in what can be summarized as localized development models. Table 3.1 presents one perspective on these variations by summarizing how implicit economic development strategies varied among the so-called four tigers of East Asia.

The table indicates that, in solving the problem of becoming internationally competitive in labour-intensive segments of manufacturing, major differences existed in state–capital relations, state–labour relations, and state–civil society relations. The emergence of large-scale oligopolies in Korea contrasts sharply, for example, with the continued small and micro-enterprise bases of industrialization in Hong Kong. Similarly, direct state use of force to regulate labour stands in Korea contrasts with indirect controls through housing and urban services provided by the state in Singapore.

As shown in the table, differences in state–society relations are also manifested in and facilitated by patterns of urbanization and investment in the built environment. Reliance on smaller-scale firms and more traditional cultural institutions to discipline labour in Taiwan has, for example, resulted in a more decentralized pattern of urbanization than in Korea. Similarly, the reliance on housing construction as a means of creating steady supplies of low-cost labour in Singapore and Hong Kong produced a very different urban setting from that of Korea, at least up to the late 1980s, where the housing crisis represented a key source of anti-statist sentiment.

TABLE 3.1. Implicit developmental models in East Asia

Relation of the state to:	Hong Kong	Korea	Singapore
Domestic capital	'Accommodationist': no explicit industrial strategy, indirect subsidies to firms via housing and infrastructure. Small-scale firms dominate.	'Corporatist state': allied with chaebol—large-scale oligopolies created/regulated by nationalization of banks.	'Extroverted corporatist': extreme reliance on TNCs and biased against domestic firms.
Transnational capital (TNCs)	Same as above	Least reliance on DFI; used military alliance with USA to influence investors.	Virtually exclusive reliance on TNCs.
Labour	Indirect control via housing, price controls, and management of international labour migration. Small-scale firms dampen worker organization.	Direct support of proletarianization and direct suppression of labour via use of police power. Chaebol and 'Fordist' factory systems foster militant labour.	Largely indirect via housing, international migration control, wage controls. Unions largely eliminated in 1960s.
Civil society	Colonial, managed by public welfare and other agencies.	Authoritarian use of police power to suppress social movements.	Overt use of state power to regulate behaviour and muzzle press in the name of 'communitarianim'.
Urbanization	Spatially compressed/ 'reflected' off China. Housing used to organize and discipline labour, urban structure to accomplish micro-enterprises.	Spatially polarized bicoastal pattern increasingly dominated by Seoul; industrial estates used to spatially organize large-scale manufacturing.	Massive reconstruction of the built environment; public housing with industrial zones for TNC investment.

Note: DFI = direct foreign investment.

The City as Multiple Landscapes

The fourth theme is that cities, as socially constructed agglomerations of human activity, contain multiple landscapes. Although from on high it may appear to be a single 'machine' in which structures and functions are systemically bound, the city is invariably composed of elements that are imperfectly integrated, endowed with different meanings, and may even contravene each other (Lawrence and Low 1990). While, for example, the overall composition of the built environment is given to the project of legitimizing the status quo and, in particular, the power of dominant groups within a given territorial formation, various segments of society may identify specific areas or sites as sanctums for their culture or subculture. Again, in the celestial city of China, access to many areas was forbidden to the majority of people, and other zones were occupied by aristocrats, retainers, military élites, and selected merchants. Thus Hangzhou, one of the world's largest cities in the thirteenth century, was dominated by the Imperial Way and the Imperial Palace yet consisted of quarters where the lives of different classes were largely played out: the rich in spacious areas to the south with summer pavilions in the hills, the poorer segments of society in the densely crowded areas off the Imperial Way.

There were also areas of common interaction among the lesser officials and the less privileged, including the merchants, traders, travellers, and the many others who kept the city alive and who met and mingled at restaurants, pleasure grounds, and huge popular theatres that spilled out from the more densely settled sectors (Gernet 1977).

Both quartered and shared spaces were not only contested but also endowed with different meanings and symbolic content. Jerusalem may represent an extreme instance of contested space, with major world religions each holding the city sacred for their respective beliefs. In contemporary India, Muslim and Hindu populations contest for the construction of temples and mosques in areas both hold sacred.

Even in religiously homogeneous societies, various strata and segments of the population give different interpretations to the same space. What is sacred to some is a symbol of oppression to others. What was officially dedicated to one type of use is perhaps just as often transformed at night or on the margins for other uses. In other words, the city has always been 'soft' (Raban 1974), never just a single city occupying a given space, but several cities, each with different meanings and uses for various classes and segments of the population.

The implications of this theme go beyond the conventional idea of 'contested space' in that it also suggests that no culture or political system is ever totalizing or completely hegemonic. Just as there are systemic forces at play, there are also likely to be anti-systemic forces that invest new meanings in existing symbolic forms or seek to create new structures in the built environment as a means to transform social relations. Mao's decision to select Beijing as the capital and to use Tiananmen as the site for proclaiming the People's Republic was a powerful use of the symbolic value of the ancient city to capture popular consciousness of China's greatness for a new epoch. Its more recent incarnation as the site for popular dissent against the state shows again how the symbolism embodied in the built environment can be used in the process of social transformation. The capture of a site may be less of a contest over its functional use than its symbolic use and the social power derived from the ability to reinterpret its symbolic meanings. Just as culture is always in a state of 'becoming', the multiple landscapes of the city are subject to continuing changes in use and symbolic interpretation as socio-cultural relations and structures change.

Not all urban space is overtly contested. Multiple landscapes can and do coexist, and while the dominant cultures and organizing principles may seek hegemony, they may, either wilfully or by default, allow other cultures and themes to be incorporated into the fabric of the city. China has cities and regions populated by Koreans and other minorities. Within cities there is more often than not a quarter set aside for foreigners, minorities, or activities desired by the ruling élites but deemed too unrespectable to be in the city proper. Japan set up *dejima* and areas of selected cities to allow for Western barbarians to carry out important trade functions. In some cases, the dominant and subordinate cultures are mixed in architectural form, as is the case of Islamic mosques in Beijing.

All of these points raise the question of just how 'orthogenetic' pre-colonial cities and centres of empires actually were.[13] Cities even in closed societies such as feudal Japan were not free from external influences; cities under the rule of even the most powerful élites still allowed for 'barbarians' and other cultural influences. This perhaps remains a matter for more empirical research, but since the relationships between society and city are themselves dynamic, such research, if it is sensitive to the view of the city by non-élites, outsiders, and the marginalized, will most likely reveal epochs of ongoing change composed of shifting, multiple landscapes rather than, for example, an 'essential' Confucianist landscape.

Societal Guidance and the Future of Cities in East Asia

If no single interpretation of the city is held in common by its residents, and no system of controlling or managing the city is totalizing, this suggests that no one vision of the city can be automatically privileged over another. This leads to one of the most striking questions in visioning the future of the urban society, namely, whose voice should be heard. Societies of East Asia have reached a juncture in history that is calling for substantial social, political, and economic restructuring to renew the basis for economic growth and higher levels of material welfare. Efforts to construct new forms of societal guidance are also focused on restructuring the city, both as the built environment for human endeavours and as an arena for the formation of a more inclusive political process.

With the possible exception of Singapore (see Ho, Ch. 11 below), the city as a functioning habitat for daily life and economic advancement is already seriously threatened by what are among the world's worst records of environmental degradation and pollution that are directly affecting human health, well-being, and productivity (UNESCAP 1993). This issue is likely increasingly to become a rallying point for the mobilization of otherwise disparate anti-statist, 'green', and even moderate reform elements as urban

---

[13] Singer and Redfield's (1954) classic exposition on the cultural role of cities divides cities into two types: the 'orthogenetic', or city characterized by a single, long-established cultural and related moral order; and the 'heterogenetic', the city of the technical order characterized by 'modern' culture that arose principally through the penetration of Western colonialism and the ascendancy of the market system.

middle classes expand and make new demands for improvements in living standards that go beyond wages and working conditions to include collective consumption and the condition of urban space.

Mixed with environmental movements will be continuing struggles for democratic reform and greater accountability of the state in its use of power and public resources: whether in the form of anti-corruption drives, as in the case of the end of the unchallenged dominance of the ruling party in Japan, or in substantial political reform, as in Korea and Taiwan, or successful reassertion of state ideologies over the expansion of political community, as in Singapore, cannot be easily predicted. But what seems to be certain is that pressures for reform will increase and that the city as habitat will be a platform, and even target, for many of these efforts.

New dimensions of the international economy are adding to the pressures for restructuring and reform. Spatially, the rapid transnationalization of indigenous firms now moving labour-intensive segments of production out of Japan, Korea, Taiwan, and Hong Kong and into other East and South-East Asian economies is fundamentally transforming the spatial and industrial structures of both sending and receiving countries. In the sending countries it has already resulted in a renewed phase of spatial polarization around capital city regions, some of which have become or are actively posturing themselves as 'world cities' (Douglass 1993a, 1993b). Secondary industrial regions in these countries, such as the southern coast of Korea or the *kansai* region of Japan, have experienced relative and even absolute economic decline and, for the first time in several decades, significant levels of structural unemployment. In Hong Kong and Singapore, it has resulted in the development of 'transborder' urban regions that, in the case of Singapore, directly extend into two neighbouring countries (Douglass 1995).

In receiving countries such as China, transnational investments are accelerating the growth of coastal metropolitan regions and generating new regional imbalances that, while perhaps being a boon to national economic growth, are mixed with ongoing economic reforms creating an urban 'floating population' numbering more than 100 million people (Linge and Forbes 1990). This is also forcing to the front of the social agenda one of the key dilemmas in the development of a market economy, namely, access to urban land and housing that, under market prices, will be vastly beyond the means of a very large proportion of the already established as well as the new urbanward migrant population.

Some analysts, viewing the simultaneous internal and external pressures on societies and states, have expressed alarm at the apparent erosion of the 'developmental' state that has been a key feature of economic growth in East Asia. Agnew and Duncan (1989), in seeking to reinsert the importance of place in social science and history, support this view by arguing the more general case that 'the power of the state has been fatally compromised by the emergence of a global system' by diminishing its capacity to regulate economic relations. This is also part of a multidimensional crisis of the state and its legitimizing ideologies that is also being propelled by the emergence of globalized systems of communications, including international television networks emerging from East Asia. Added to this is the increasing commodification of human relationships and exchange that subordinates social and cultural institutions to the utility of the market.

Whether or not the new pressures on the state should be lamented or applauded is a matter of considerable debate. The dismal human rights records of most governments in East Asia, including the brutal suppression of civil society in some, cannot be cloaked in the guise of cultural integrity or the importance of 'place'. At the same time, however, political reform should not be equated or confused with a diminishment of the scope for action in the public domain. In many instances this scope needs to be enlarged if solutions are to be sought for contemporary social issues. Nowhere is this more prominent than in East Asia where economic change and new patterns of spatial polarization have led to widening social inequalities, severe environmental deterioration, and other major sources of social discontent that are not easily amenable to market solutions.

In a more positive vein, the question to be asked is to what extent can new forms of societal guidance (Friedmann 1987) be created through new or transformed socio-cultural and political institutions? These new forms, in Lefebvre's (1991: 59) vocabulary, will in turn call for the production of an appropriate urban space. In this context, the logic of the framework developed in this chapter is that, despite a shared set of cultural traits, there is no single East Asian model of societal guidance. Variations in past histories, the contemporary state, and society reveal significantly different outcomes even in the process of export-oriented industrialization followed in Japan, Korea, Taiwan, Hong Kong, Singapore, and, more recently, China.

The dilemmas faced by these societies are none the less similar in many ways. One common to most is how to counterbalance increasing concentrations of population in a few mega-urban regions where pressures on inherited urban form are intense and are manifested in deteriorating environments and heightened social cleavages related to access to land, housing, urban infrastructure, and other elements of the built environment such as public space and urban services. All of this is taking place in an era in which the awakening of civil society is also very much part of changing social relations and agendas.

While much needs to be said in defence of human agency and the prospects of individuals and collectivities of people to make history (Giddens 1984; Friedmann 1989), it is not certain that this history can ever be very far removed from, in the case of the current moment, systemic global forces that are generating many of the pressures impinging on East Asian cities.

The phenomenal collapse of space–time relations made possible by revolutionary advances in transportation and communications and the emergence of corporate entities capable of taking advantage of them on a global scale have made these pressures all the more invasive as well as pervasive. Whether, as some have argued, this will lead to a 'denationalization' of history (Agnew and Duncan 1989) and a 'rendering of the world as a single place' (King 1989: 5), or, to the contrary, new social and political structures will emerge to strengthen local capacities to guide economic forces, remains to be seen.[14] Whether, too, the city as a principal node of globalized consumer markets will be the locus for increasing global convergence of sociocultural relations and urban patterns of development, or will instead become the site for resistance to commodification and new pressures for asserting life space over economic space (Friedmann 1988), is also a question without a clearly evident answer. In either case, however, a new period of social change and spatial restructuring has begun. This is itself a climacteric event in East Asian history.

Conclusions

Social change is the outcome of shifting constellations of power—including external as well as internal forces—comprising the political process of each society. Urbanization has been an essential dimension of social change throughout the world, and cities are also transformed with social change measure by measure. The urban transition experienced in East Asia in the latter half of the twentieth century represents one of the most profound processes of socio-cultural, political, and economic change in the history of these societies. Great cities, which were the centres of empires and tributary systems, have been part of East Asia for millennia. But in the light of contemporary times, they were few in number, relatively small in size, extended control over essentially agrarian economies, and controlled hinterlands in a much more rudimentary way than modern communications and transportation technologies allow. Many have subsequently been overshadowed by coastal cities created in colonial times that have continued their ascendance as centres of production and trade for world markets.[15]

---

[14] King (1989), for example, argues that 'homogenisation of space, associated with the urban effects of the new international division of labour and the internationalisation of capital over the past two decades, has led to increasing convergence between the characteristics of the so-called first and third world cities'.

[15] Whereas Tokyo was the first city in the world to reach a population size of 1 million in the 19th cent., East Asia now has metropolitan regions reaching 30 million inhabitants (Tokyo) with some accounting for almost half of their national populations (Seoul). Earlier projections for the year 2000 show eight East Asian cities with more than 5 million in population (Tokyo, Shanghai, Beijing, Seoul, Osaka, Tianjin, Wuhan, and Hong Kong). These estimates severely underestimate the actual built-up area of these cities, which extends well beyond their official administrative boundaries (Douglass 1989).

National urban systems have also developed along industrial corridors to create increasingly complex networks of interaction.

In the cities making up these networks, the imperial palace has yielded to the presidential palace, the temple to the bank, the quarters of the hereditary aristocracy to the condominiums of the nouveaux riches, and the city gate to the expressway toll booth. An urban middle class with aspirations to a greater voice in governance has emerged where none existed before. In far greater number are the members of the urban working class that was also created in large part over the past four decades. Within this class are also various segments ranging from semi-protected union workers to the most vulnerable day labourers. In sum, society and the city have been vastly transformed in a matter of just a few decades.

Although no longer the dominant forces defining social relations, the socio-cultural institutions and their manifestations in older layers of the built environment of the city still exist. These are not simply relics of the past, but serve vital roles in the social, political, and economic life of cities. As sets of values, even cultural histories muted by time are still called upon to legitimize holders as well as seekers of power. As part of urban form they are emblems of cultural continuity that serve to provide the occasion for national celebrations, as exemplified by the recent celebration of the 600th anniversary of Seoul, consciously to identify a Korean, Japanese, or Chinese way of life, and even as sites to rally anti-statist demonstrations in the name of a larger historical legitimacy and mission.

The layering of history is a localized process. The socio-cultural, political, and economic dynamics focusing on Korea's cities are not the same as those of Japanese or Chinese cities, and key differentiating characteristics are related to cultural and social differences that have historically developed, overlaid, and transmuted over time in each. As in the past, the future of cities in East Asia is also contingent upon the dynamics of socio-political continuity and change within each social formation.

East Asian societies are each now facing an unusually critical moment. Although Hong Kong may present an interesting exception, the reliance on labour-intensive forms of manufacturing, at least in the cases of Korea, Taipei, and Singapore, is no longer viable. The state apparatus used to maintain supplies of cheap industrial labour has, in this sense, become anachronistic. At the same time, the authoritarian, highly centralized regimes of the post-Second World War era are no longer able to legitimize their totalizing forms of control in the name of anti-communism, and pressures for creating a new basis in societal guidance through democratic institutions are likely to continue.

In this context, the longer-term sustainability of cities will rest on the capacity to generate new socio-cultural institutions capable of galvanizing collective energies to reverse many of the trends appearing over the past few decades, including massive urban environmental deterioration, reliance on unprotected urban labour, mostly women, to prop up export drives, general suppression of civil society, and, particularly since the mid-1980s, rising social inequalities. Just as the past four decades have revealed many pathways toward export-oriented urban industrialization in East Asia, they have shown that many nations, notably in Latin America, have been unable to sustain their 'newly industrializing country' status in a manner that can genuinely be called developmental. In the USA and Europe as well, cities that were once seen as vital engines of industrial growth have become exhibitions of social dislocation and the loss of collective will to confront it.

Whether a new era of development can emerge from the conjuncture of social, political, and economic pressures for change is a question that is not only appearing in East Asia but in North America and Europe as well. The view taken here is that the array of possibilities for reconstructing the foundations for new phases of development at critical junctures in history are greater than received world systems theories allow, but also more difficult to achieve than the conventional models of international diffusion of modernization can comprehend. No matter how difficult or easy, the future of East Asian cities will be determined as much by localized socio-cultural and political processes as by global imperatives.

References

Abu-Lughod, Janet (1991). *Changing Cities*. New York: Harper Collins.

Agnew, John A., and Duncan, James S. (1989). Introduction, in John A. Agnew and James S. Duncan (eds.), *The Power of Place: Bringing Together Geographical and Sociological Imaginations*. Boston: Unwin Hyman.

Amsden, Alice H. (1989). *Asia's Next Giant: South Korea and Late Industrialization*. New York: Oxford University Press.

Apter, David, and Sawa, Nagayo (1984). *Against the State*. Cambridge, Mass.: Harvard University Press.

Armstrong, Warrick, and McGee, T. G. (1985). *Theatres of Accumulation*. London: Methuen.

Basu, D. (ed.) (1985). *The Rise and Growth of Colonial Port Cities in Asia*. Lanham, Md.: University Press of America.

Carter, Harold (1983). *An Introduction to Urban Historical Geography*. Baltimore: E. Arnold.

Castells, Manuel (1977). *The Urban Question*. London: Edward Arnold.

Corbridge, Stuart (1986). *Capitalist World Development: A Critique of Radical Development Geography*. Totowa, NJ: Rowman & Littlefield.

Cotton, James (1992). 'Understanding the State in South Korea: Bureaucratic-Authoritarianism or State Autonomy Theory?', *Comparative Political Studies*, 24 (4): 512–31.

Deyo, Frederick (1990). 'Economic Policy and the Popular Sector', in Gary Gereffi and Donald Wyman (eds.), *Manufacturing Miracle: Paths of Industrialization in Latin America and East Asia*. Princeton: Princeton University Press.

Douglass, Mike (1989). 'The Future of Cities on the Pacific Rim', *Comparative Urban and Community Research*, edition on *Pacific Rim Cities in the World Economy*, 2: 3–58.

——(1991). 'Transnational Capital and the Social Construction of Comparative Advantage in Southeast Asia', *Southeast Asian Journal of Social Science*, 19 (1–2): 14–43.

——(1992). 'Global Opportunities and Local Challenges for Regional Economies', *Regional Development Dialogue*, Summer (13): 3–21.

——(1993*a*). 'Socio-political and Spatial Dimensions of Korean Industrial Transformation', *Journal of Contemporary Asia*, 23 (2): 149–72.

——(1993*b*). 'The "New" Tokyo Story: Restructuring Space and the Struggle for Place in a World City', in Kuniko Fujita and Richard Child Hill (eds.), *Japanese Cities in the Global Economy: Global Restructuring and Urban-Industrial Change*. Philadelphia: Temple University Press.

——(1994). 'The "Developmental State" and the Asian Newly Industrialized Economies', *Environment and Planning*, 26 (A): 543–66.

——(1995). 'Global Interdependence and Urbanization: Planning for the Bangkok Mega-urban Region', in T. G. McGee and Ira Robinson (eds.), *The New Southeast Asia: Managing the Mega-urban Regions*. Vancouver: University of British Columbia Press.

Dunford, M. (1990). 'Theories of Regulation', *Environment and Planning*, 8 (D): 297–321.

Esser, J., and Hirsch, J. (1989). 'The Crisis of Fordism and the Dimensions of a "Postfordist Regional and Urban Structure"', *International Journal of Urban and Regional Research*, 13: 417–37.

Featherstone, Mike (1991). *Consumer, Culture and Postmodernism*. Newbury Park, Calif.: Sage Publications.

Friedmann, John (1987). *Planning in the Public Domain: From Knowledge to Action*. Princeton: Princeton University Press.

——(1988). *Life Space and Economic Space: Essays in Third World Planning*. New Brunswick, NJ: Transaction Books.

——(1989). 'Collective Self-Empowerment and Social Change', *Ifda Dossier*, 69: 3–14.

Gernett, Jacques (1977). 'Daily Life in China on the Eve of the Mongol Invasion 1250–1276', in Janet Abu-Lughod and Richard Hay, Jr. (eds.), *The World Urbanization*. New York: Methuen.

Giddens, A. (1984). *The Constitution of Society: Outline of the Theory of Structuration*. Berkeley and Los Angeles: University of California Press.

Ginsburg, Norton, Koppel, Bruce, and McGee, T. G. (eds.) (1991). *The Extended Metropolis: Settlement Transition in Asia*. Honolulu: University of Hawaii Press.

Han, Woo-keun (1974). *The History of Korea*. Honolulu: University of Hawaii Press.

Hsiao, H. H. M. (1990). 'An East Asian Development Model: Empirical Explorations', in P. Berger and H. M. Hsiao (eds.), *In Search of an East Asian Development Model*. 2nd edn. New Brunswick, NJ: Transactions Books.

Johnson, E. A. J. (1970). *The Organization of Space in Developing Countries.* Cambridge, Mass.: Harvard University Press.

King, Anthony D. (1984). 'The Social Production of Building Form: Theory and Research', *Environment and Planning*, 2 (D): 429–46.

——(1989). 'Colonialism, Urbanism, and the Capitalist World Economy', *International Journal of Urban and Regional Research*, 3: 1–18.

Lawrence, Denise, and Low, Setha (1990). 'The Built Environment and Spatial Form', *Annual Review of Anthropology*, 19: 453–505.

Lefebvre, Henri (1991). *The Production of Space.* Oxford: Basil Blackwell.

Linge, G. J. R., and Forbes, D. K. (eds.) (1990). *China's Spatial Economy.* Hong Kong: Oxford University Press.

Low, S. (1988). 'Cultural Aspects of Design', *Architectural Behavior*, 4: 187–95.

——(1990). 'Urban Public Spaces as Reflections of Culture: The Plaza in Costa Rica', in L. Duhl (ed.), *Urban Condition II.* Beverly Hills Calif.: Sage.

McGee, T. G. (1967). *The Southeast Asian City.* London: Bell.

McGuire, R. H., and Schiffer, M. B. (1983). 'A Theory of Architectural Design', *Journal of Anthropological Archaeology*, 2: 227–303.

Moody, P. (1988). *Political Opposition in Post-Confucianist Society.* New York: Praeger.

Morishima, M. (1982). *Why has Japan 'Succeeded'?* Cambridge: Cambridge University Press.

Murphey, Rhoads (1984). 'City as a Mirror of Society: China, Tradition, and Transformation', in John A. Agnew, John Mercer, and David E. Sopher (eds.), *The City in Cultural Context.* Boston: Allen & Unwin.

O'Donnell, G., Schmitter, P., and Whitehead, L. (eds.) (1986). *Transitions from Authoritarian Rule: Prospects for Democracy.* Baltimore: Johns Hopkins University Press.

O'Malley, W. (1988). 'Culture and Industrialization', in Helen Hughes (ed.), *Achieving Industrialization in East Asia.* Cambridge: Cambridge University Press.

Pred, Allan (1984). 'Structuration, Biography Formation, and Knowledge: Observations on Port Growth during the Late Mercantile Period', *Environment and Planning*, 2 (D): 251–75.

——(1986). *Place, Practice, and Structure: Social and Spatial Transformation in Southern Sweden, 1780–1850.* Totowa, NJ: Barnes & Noble.

Raban, Jonathan (1974). *Soft City.* London: Fontana/Collins.

Rapoport, Amos (1984). 'Culture and Urban Order', in John A. Agnew, John Mercer, and David E. Sopher (eds.), *The City in Cultural Context.* Boston: Allen & Unwin.

Redfield, R., and Singer, M. S. (1954). 'The Cultural Role of Cities', *Economic Development and Cultural Change*, 3: 53–73.

Rey, Pierre-Philippe (1971). *Colonialisme, néo-colonialisme et transition au capitalisme.* Paris: Maspéro.

Robertson, R. (1988). 'The Sociological Significance of Culture: Some General Considerations', *Theory, Culture, and Society*, 5: 3–23.

Samuels, M., and Samuels, C. (1989). 'Beijing and the Power of Place in Modern China', in John A. Agnew and James S. Duncan (eds.), *The Power of Place: Bringing Together Geographical and Sociological Imaginations.* Boston: Unwin Hyman.

UNESCAP (United Nations Economic and Social Commission for Asia and the Pacific) (1993). *State of Urbanization in Asia and the Pacific, 1993*. Bangkok: UNESCAP Division of Industry, Human Settlements, and Environment.

Vogel, E. F. (1991). *The Four Little Dragons: The Spread of Industrialization in East Asia*. Cambridge, Mass.: Harvard University Press.

Wallerstein, Immanuel (1979). *The Capitalist World System*. Cambridge: Cambridge University Press.

Walton, John (1984). 'Culture and Economy in the Shaping of Urban Life: General Issues and Latin American Examples', in John A. Agnew, John Mercer, and David E. Sopher (eds.), *The City in Cultural Context*. Boston: Allen & Unwin.

4

Work, Space, and Place in the Cities of the East Asian Pacific Rim

T. G. McGee and Gisèle Yasmeen

> Each urban tradition like every city, represents a continuing accommo-
> dation of global societal tendencies to particular sets of demographic
> and environmental exigencies.
>
> (Lampard 1965: 519)

Introduction: East Asia and the Urban Transition

The approach of this volume, to quote the editors, 'is to expertly deal with
the cultural and historical dimensions of the urbanization process in Asian
cities differently from the dominant approach in development studies which
treats cultural and historical elements as residual factors'. This dimension of
urbanization that the editors request poses perhaps the greatest challenge.
Cities are almost invariably seen as the major institutional component of
the economy of all societies.[1] Thus, it is no easy task to relegate economic
functions to a residual role. Especially so in an essay that is concerned with
changing economic functions and work patterns in Asian cities. Finally, the
urban place is such a complex institution that the task of separating the
economic, social, political, and cultural features is almost impossible.

The countries of 'East Asia' as defined by this volume are variously
engaged in a process of socio-cultural and demographic transformation
known as the 'urban transition'. Conventionally, the urban transition is
measured by the proportion of a given spatial unit (generally the nation-
state) resident in urban places at different points in time. This is often
befuddled by different definitions of what is 'urban'.[2] We will take a simple

[1] By this statement we do not wish to suggest that cities do not also perform important
political and cultural roles in society, simply, that their economic functions are universal. Even
pilgrimage cities such as Varanasi (Benares), Lourdes, Jedda, and the Vatican City are performing important economic roles.
[2] Urban definitions vary widely within the East Asian Pacific Rim. Thus, in some countries,
demographic definitions are utilized (Indonesia), while in others political definitions are used
(such as in Malaysia).

approach and assume that a country becomes urbanized when at least 50 per cent of its population is resident in urban places. The economic and political importance of cities in less urbanized societies, however, must continually be taken into account in addition to demographic factors.

Whereas the 'West' experienced the urban transition largely during the nineteenth century covering a span of at least 100 years (Williamson 1988), the so-called 'Third World' has been urbanizing since the Second World War at a far more rapid rate. The shift is also of greater demographic importance. Hundreds of millions of people are involved in the shift from an agrarian to an urban way of life. This contrasts sharply with the Western urban transition which directly affected far fewer numbers. Differences of opinion about the effect of these processes of economic growth have been the subject of ongoing debate (see Armstrong and McGee 1985; McGee 1967, 1971, 1995).

Asia (defined here to include East, South, South-East, and Western Asia)[3] will account for almost 61 per cent of the world's population and 38 per cent (1.2 billion) of the world's urban population by the year 2005 (UNESCAP 1993). This places major challenges to the ability of Asian governments to create viable employment opportunities for their numerous urbanites as well as provide adequate infrastructure for the booming population. The purpose of this chapter is to compare and contrast the experiences of various East Asian cities in the realm of work. Work is here defined broadly as a process whereby labour is carried out to produce some form of value. We deliberately make room for the world of unpaid and 'informal' labour—much of which is often carried out by women and the urban underclass—which is often neglected in theoretical and empirical discussions of work.

It is understood that work is deeply embedded in the concept of culture. Indeed it is central to the Western-derived theories of the rural–urban difference continuum which attempt to explain social, economic, and political behaviour (see McGee 1971). The authors, however, are highly critical of this theory because of its Eurocentric origins, its limited attention to gender differences, and the more nuanced relationship of work, space, and place in Asian cities.

The specific purpose of this chapter is to shed light on the interrelationships between work, space, and place in the more or less cohesive cultural realm of 'East Asia'. Indeed, the ostensible cultural similarities between the various East Asian countries is a matter of debate (see Douglass and Kim, Ch. 12 below). In order to understand properly what may be unique

---

[3] The entire definition of 'Asia' as a region has been a matter of debate and vehement criticism over the past fifteen years (Said 1979). No equivalent term exists in any 'Asian' language. Indeed, the notion of 'Asia', as opposed to 'Europe', is a concept which has greater relevance to the historiography of conquest than to bona fide regional geography (see McGee 1991).

configurations of work, space, and place in this 'region', it is necessary to make cursory comparisons with the experiences of East Asia's neighbours (South and South-East Asia) as well as that of the 'West'.

The Urban Transition in East Asia

The urban transition in East Asia is proceeding at variable rates throughout the region. Of course, this is not to argue that the urbanization process and economic growth have occurred simultaneously throughout the region. Table 4.1 shows that these processes have not occurred evenly. This unevenness is indicated by grouping the countries of the East Asian Pacific Rim into three categories. First, the core industrialized countries which have levels of urbanization of 75 per cent or more. Secondly, those countries in which the urban transition is in process, which include states in which industrialization is proceeding very rapidly such as Malaysia and Thailand; and, finally, the two socialist economies, namely China and Vietnam, which are currently undergoing transition to forms of mixed-market economies.

To some extent this pattern of regional urbanization fits the 'geese flying south' model with Japan leading the way and Indonesia, Thailand, China, and Vietnam being the last landing places. This is further supported by economic data assembled by Arrighi, Ikeda, and Irwan (1993) which involve a careful analysis of the East Asian Pacific Rim economic miracle. They comment that 'The East Asian miracle is first and foremost a Japanese miracle' (Arrighi *et al.* 1993: 42) with the other components experiencing

TABLE 4.1. City types

Characteristics	Pre-industrial	Industrial	Post-industrial
Technology	Animate power	Inanmimate power	Electronic
Social structure	Small élite in power; mass population of artisans, servants, outcastes.	Capitalist élite; industrial working and middle classes; mass education.	New Middle Class; fragile underclass in menial labour; demise of state and unions.
Spatial structure	Symbolic city centre; large markets; area specialization; shop-houses.	CBD; factories in centre with worker housing; suburbs for middle class.	Gentrification; deindustrialization of West; Third world EMR? Consumption.

Note: CBD = commercial business district; EMR = extended metropolitan region.
Sources: Sjoberg (1968); Agnew *et al.* (1984).

later economic growth which still proportionally is much less significant than that of Japan. This raises important questions as to the historical importance of Japan in the region and its effect on urbanization patterns (see Dirlek 1993 and Palat 1993).

It also raises important issues relating to the reasons why this region has performed, and is performing, so well economically at a global scale. This has been the catalyst for another debate that is important to our discussions as well as the role of 'culture'. 'Culture', as Raymond Williams explained in *Keywords* (1976), is one of the most difficult words in the English language. We are using the term to refer to a socially constructed and negotiable system of shared meaning and practices. On one side, a group of writers have attempted to explain the success of the East Asian model as occurring because of the existence of a Confucian set of values that include respect for family, deference to authority, trust, discipline, and the virtue of hard work. In itself this cluster of values is not unlike the so-called 'Protestant ethic' that helped spawn Western capitalism and industrialization (Morishima 1982; Vogel 1991) although in both cases over-generalizations are often made based on stereotypes which are not valid at the level of locally lived experience.

Certainly along with the arguments concerning the Confucian quality of the 'East Asian' way there appears to be a reputation in the region for more disciplined labour forces, weaker unions, and more consensual relations in worker–management matters compared to that which exists in developed Western market societies. It is often argued that these elements contribute to a different 'culture' of work in the East Asian Pacific Rim.[4]

While this argument was originally directed at the first five NICs (Japan, Korea, Taiwan, Singapore, and Hong Kong), it is now extended to China and Vietnam (socialist), Thailand (Buddhist), and Malaysia and Indonesia (Islamic), which seems perplexing. However the protagonists of this view accommodate this contradiction by stressing the dominant role of the overseas Chinese in these countries operating a form of Chinese capitalism in the pores of the dominant economic system. Thus, if Taiwan and Hong Kong are included, Redding estimates that for the overseas Chinese, 'taking account of their officially unacknowledged power in the ASEAN economies, the wealth generated per year would be somewhere in the region of US \$200 billion GNP with around 40 million people' (Redding 1993: 3). These statistics support the view advanced by Yoshihara (1988) that the overseas Chinese play a major role in the economy of all South-East Asian countries. This introduces the idea that there is a form of 'ethnic space' which operates in networks across borders, what Arjun Appadurai termed an 'ethnoscape' (Appadurai 1990). The East Asian Pacific Rim is indeed a

---

[4] We have not taken up the theme of labour–management relations in the East Asian Pacific region in this chapter. There are many studies in the 'business' literature that deal with this aspect of work.

highly culturally diverse region with respect to the work-related values as this chapter will attempt to illustrate.

On the other side are the writers who interpret the region's progress from a structural perspective, arguing that the economic successes of countries in the East Asian Pacific Rim are due to a combination of factors such as the existence of a stable rule by a political bureaucratic élite with developmental goals, co-operation between private and public sectors, the existence of a cheap, disciplined, and well-educated labour force, the fortuitous conditions of the Cold War, and access to global markets.[5] Further, patterns of global economic restructuring in the past twenty years are used as explanations for the deindustrialization of the Western 'core' and the industrialization of the former geographical 'periphery', that is, countries of the Pacific Rim (excluding Japan).

In fact, it may be argued that neither of these two polarized positions offers a satisfying explanation for the characteristics of the urban labour markets of Asia. Some parts of Asia, such an India, Bangladesh, Indonesia, and Vietnam, continue to suffer underemployment and others such as Japan, Singapore, and Hong Kong have proceeded so far in the creation of viable productive employment opportunities that they are now experiencing labour shortages. If one is concerned, however, with tracing the development of the labour markets of urban areas in the most rapidly industrializing countries of Asia (the focus of much of this volume), the most useful model may be that of Frederick C. Deyo (1989), who introduces the concept of 'Labour systems'. These labour systems 'define the means through which labour is transformed into products (including services) and more liquid economic assets (e.g. wages, benefits, profits)' (Deyo 1989: 11). This concept emphasizes the role of changing 'labour market conditions or political structures, social welfare programs, autonomous non-elite subsistence strategies, struggle in labour market shelters' behind which 'alternate systems of labour expropriation' (Deyo 1989: 13) are carried out. In many East Asian economies, institutions of patriarchy, communal paternalism, bureaucratic paternalism, and patrimonalism affect usually all aspects of the labour system ranging from recruitment to retirement.

From the point of view of this essay, we believe that it is possible to see these two explanations operating together rather than in an opposing fashion. Thus the economic growth of the region and its growing urbanization have been the result of what has been described elsewhere as the operation of three circuits of capital. First, flows of capital from outside the region; secondly, flows of capital from within the region in which Japan, Korea, and the overseas Chinese are representative; and finally indigenous capital generated at the country level used for domestic investment. 'Circuits of capital' have operated together to provide the capital for labour

---

[5] The arguments for this portion are set forth in Vogel (1991).

force growth and the creation of the built environment of the rapidly urbanizing mega-cities of the region.[6] Certainly if one accepts the idea of urban labour force transformations of these groups of societies being mediated by culture defined as a system of shared values and practices, these two positions are easily bridged and some legitimate coherence of the region accepted.

At a global level, the processes of labour formation in urban areas have been influenced by processes of economic restructuring, international investment, the collapse of socialist societies, and technological advances in transportation and communication which have differential impacts on various components of the urban system of developing countries. As many commentators have indicated, earlier interpretations of this process, which presented labour transformation as having what Deyo calls 'a common set of social consequences' (Deyo 1989: 10), have been subject to considerable modification. Not least is the assumption that global imports produce a common local response. As the work of Appadurai (1990: 6) argues, the new global order must 'be seen as a complex, overlapping, disjunctive order, which cannot any longer be understood in terms of existing centre-periphery models'. Thus, it is in the large mega-urban regions which are most heavily integrated into the international system that the labour force has undergone the most complex set of changes in terms of the global–local dialectic. This is particularly true in the case of the 'miracle' economies of East Asia and the Pacific Rim where mega-urbanization has resulted in the integration of large segments of the labour force into the international system. In the words of Nonini (1993: 162) there are 'a set of economic, political and social processes creating relationships within an area—perhaps one could call it a "supra-region" that link up different regions and that the flows of capital, people, commodities and information are being primarily articulated through the major cities of the region'.

Linking Work, Space, and Place to the Urban Transition

In order to understand the changing relationship between work, space, and place as they evolve during the urban transition in East Asian Pacific Rim cities, it is necessary to develop a conceptual model which describes the overarching framework for this analysis. While we want to use a labour market model for the analysis of the changing relationship between work, space, and place, it is important to emphasize that the 'space' and 'place' in which 'work' is embedded is constantly being altered by a series of what David Harvey (1989) following Lefebvre (1974) calls 'spatial practices'.

---

[6] The operation of the overseas Chinese capital circuit is described in Wu and Wu (1980) and McGee (1985).

Thus as the urban transition proceeds the relationships between work, place, and space are in a constant state of flux. Historically this process has been simplified by the use of 'normative' models of what are assumed to be broadly similar types of cities. This approach was first made explicit in Sjoberg's model of the pre-industrial city contrasted with the industrial city (Sjoberg 1968). More recently, discussions revolve around the appearance of the post-industrial city following periods of global economic restructuring and the collapse of space–time barriers due to advances in communication technology. Table 4.1 summarizes the characteristics of these three types of cities which to a certain extent bridge the gap between Western and Asian urbanization. Needless to say, the models of the pre-industrial and industrial cities have been heavily criticized. Wheatley (1971) has argued very strongly for the role of religico-cosmological beliefs in the pre-industrial city and others have pointed to the empirical weaknesses of the model (see Skinner 1977). It has also been pointed out that the idea of the pre-industrial city took very little account of the emergence of cities in the context of colonialism which led to a particular type of synthesized model in the nineteenth and twentieth centuries which occurred all over the 'colonial world' (see McGee 1967; King 1990). Finally many writers have indicated that the model of the industrial city is increasingly being outmoded as the role of services in many cities increases with structural change (see Agnew *et al.* 1984). This is producing a new post-industrial city in which the production base of the city is being eroded and replaced by a built environment oriented towards consumption.

Given the long historical tradition of urbanization in Asia one may find all elements of these models coexisting within many Asian cities most obviously in the built environment but also in the social structure, labour markets, and virtually every aspect of Asian life. This can be best exemplified by the image of a Hui Muslim running a foodstall selling Hezhou beef and hand-pulled noodles located in front of a shopping mall which is a joint venture between the Beijing municipality and a foreign investor. Fashionably dressed shoppers are sitting on stools eating the noodles. How do you explain the juxtaposition of these various layers of history in the life and work of Asian cities? Donna Haraway has manipulated the concept of the local–global dialectic by pointing to how, today, 'the forces of extreme localisation, of the intimately personal and individualised body, vibrate in the same field with global high tension emissions' (Haraway 1991: 195 cited by McKay 1994).

In such a complex setting of local–global relations Deyo's concept of labour systems provides the most resonance, giving a subtle understanding to the labour systems of the urban areas of the East Asian Pacific Rim. This model is also extremely helpful in the comparison of the labour force of different cities as well as the processes that are experiencing changes in their labour force structure. What is more, it offers the opportunity to accommodate the changes occurring in the 'transforming socialist economies' of the East Asian Pacific Rim as well as the market economies. In the urban areas of the East Asian Pacific Rim there is ample evidence to support the view that complex labour systems are evolving. This is occurring in a number of ways.

First, by increasing employment within the governmental sector, the recruitment of labour into various parts of government (such as the civil service and armed forces) is a major feature of the East Asian Pacific Rim, and offers an important avenue for upward mobility. This is, at least in part, due to the greater degree of involvement of East Asian Pacific Rim states in the development process than was the case in the West. Second, the marked increase in industrialization, which began with export-oriented industrialization and import substitution, has created considerable formal employment. Finally, as economic growth has continued, the formal service sector has grown with employment in occupations such as office work and communications.

At the same time in some parts of Asia, non-proletarianized labour forces have remained remarkably persistent, growing in size in the socialist cities of China and Vietnam as the economy becomes more market driven. For example, the population engaged in the informal sector of Beijing has grown from approximately 800 in 1977 to 333,000 today (Ho Bi Liang, personal communication). This labour force also remains persistent in the later arrivals on the 'development scene' such as the Philippines and Indonesia. In most urban areas the retailing sector is dominated numerically by employment in the informal sector. As this model of the urban labour market shows, these various labour 'fragments' or sectors are capable of rapid fluctuation in numbers reflecting the operation of business cycles, government decisions, and cultural practices. This disputes conventional modernization theory which postulates that the informal sector will eventually be replaced by formal-sector employment following 'development'. Deyo's concepts of 'labour systems' show that the operation of the national state, business cycles, and culture inherently influence patterns of work. This is a much more subtle view of the formation of labour markets.

Evolving Patterns of Work and the Urban Transition: An Empirical Analysis

The preceding sections make it clear that the urban transition is accompanied by significant changes in the patterns of work. Table 4.2 and Table 4.3 show clearly that as the urban transition has been proceeding the proportion of employment in agriculture is declining rapidly. Thus, in the urbanized societies of Singapore, Hong Kong, Republic of Korea, Japan, and

TABLE 4.2. Total country population and per cent urban 1970–1990

	1970		1980		1990	
	000s	% urban	000s	% urban	000s	% urban
Urbanized countries						
Singapore	2,075	100.0	2,414	100.0	2,723	100.0
Hong Kong	3,959	89.7	5,039	91.6	5,705	94.1
Rep. of Korea	31,446	50.1	37,436	68.8	43,441	79.8
Japan	104,331	71.2	116,807	76.2	123,460	77.0
Taiwan	14,675	34.0	17,805	52.0	20,353	74.0
Urban transition in process						
Malaysia	10,852	27.0	13,763	34.6	17,567	43.0
Philippines	37,540	33.0	48,317	37.4	62,413	42.6
Indonesia	120,280	17.1	146,776	22.4	179,321	30.9
Thailand[a]	35,745	13.3	46,718	17.3	55,702	22.6
Incipient urban transition						
China[b]	830,675	17.4	996,134	19.6	1,139,060	33.4
Vietnam	42,729	18.3	53,700	19.3	66,693	21.9

[a] Thailand figures are underestimated (should add 8% for under-enumeration). See Greenberg (1994).
[b] China figures overestimate figures for non-agricultural population in urban areas.

Sources: UN (1992); and other national statistical sources.

TABLE 4.3. East Asia Pacific Rim: structure of employment (% in agriculture)

	1960	1970	1980	1990
Urbanized countries				
Singapore	8.0	3.0	1.0	0.5
Hong Kong	7.0	4.0	2.0	0.2
Rep. of Korea	61.0	50.0	35.0	20.0
Japan	33.0	19.0	11.0	8.0
Taiwan	56.0	36.7	19.5	12.8
Urban transition in progress				
Malaysia	63.0	—	37.0	31.0
Philippines	59.0	54.0	51.0	45.0
Indonesia	68.0	64.0	55.0	56.0
Thailand	82.0	79.0	71.0	64.0
Incipient urban transition				
China	83.0	78.0	74.0	61.0
Vietnam	—	—	73.0	70.0

Sources: UNESCAP (1991: 2.31); Vietnam GSO (1990).

Taiwan there are now only small proportions of the population engaged in agriculture and the non-city states such as Korea, Taiwan, and Japan have seen a very rapid fall in the period since 1960, while at the same time their level of urbanization has risen to almost 74 per cent. In the states where the urban transition is well advanced but not yet complete there is a fall-off but only in Malaysia has the proportion fallen below 45 per cent. The socialist states still have almost two-thirds of their population engaged in agriculture. Paralleling this decline of agriculture has been an increase in manufacturing and industry throughout the region, most markedly in the East Asian core countries but also in South-East Asia (Table 4.4). Growth in the service sector is most remarkable in Korea, Japan, and Malaysia. In the case of Japan, Hong Kong, Malaysia, and Korea, this growth of the service industry is related to structural change as part of the process of economic growth but in the case of the Philippines, and to some extent Indonesia, it may be an expansion of low-income service activities due to the inability of the formal sector to absorb sufficient employment (Table 4.5).[7]

The growth of urbanization has been associated with a number of trends in the labour markets which are affecting the life of the cities; these in part reflect the intervening role of culture and ideology as manifested in demographic change.

The first trend relates to the pattern of population growth in the region.[8] Generally, the East Asian Pacific Rim has shown relatively modest rates of population growth compared to the rest of the developed world, reflecting that over the last twenty years fertility has fallen, reducing annual growth rates. Thus in the period between 1955 and 1990, the average annual growth rate of Latin America—a more highly urbanized continent—fell from 2.76 per cent to 2.09 per cent while that of East Asia declined from 1.53 per cent to 1.31 per cent. There were, however, substantial differences between the more industrialized countries such as Japan, Korea, Hong Kong, Taiwan, and Singapore and other parts of the region. Indonesia, Thailand, and China all exhibited dramatic decreases but the rates of growth remained high in Malaysia, the Philippines, and Vietnam.

The decline in fertility has also been associated with increased longevity as a result of improved health and standards of living. This means that there are growing numbers in the age group over 65 years. This is even more marked in the urban areas (UNESCAP 1993). When these trends are associated with the increase in the number of years of education (much higher in urban areas) the demographic impact on the labour force is considerable.

This is well illustrated by the changing patterns of labour force participation among the urban populations of the region. First, as the general population is now healthier, they are living longer and remaining in the labour force longer. At the same time, the new entrants to the labour force are entering later after completing education. There has been a decline in the labour participation rates for youths in the industrialized countries (Hong

---

[7] However, in some informal-sector occupations, especially vending, wages are sometimes higher than in the formal sector (McGee and Yeung 1977; Tinker 1987).

[8] This section is based on Neville (1993).

TABLE 4.4. East Asia Pacific Rim: structure of employment (% in manufacturing and industry)

	1960	1970	1980	1990
Urbanized countries				
Singapore	21.0	30.0	35.0	36.0
Hong Kong	50.0	54.0	50.0	31.0
Rep. of Korea	8.0	20.0	27.0	35.0
Japan	31.0	34.0	33.0	34.0
Taiwan	—	20.9	32.7	31.2
Urban transition in progress				
Malaysia	—	—	24.0	25.0
Philippines	15.0	17.0	16.0	16.0
Indonesia	7.0	8.0	13.0	15.0
Thailand	4.0	5.0	10.0	11.0
Incipient urban transition				
China	6.0	10.0	16.0	17.0
Vietnam	—	—	15.0	15.0

Sources: UNESCAP (1991: 2.31); Vietnam GSO (1990).

TABLE 4.5. East Asia Pacific Rim: structure of employment (% in services)

	1960	1970	1980	1990
Urbanized countries				
Singapore	71.0	66.0	63.0	63.0
Hong Kong	43.0	42.0	48.0	63.0
Rep. of Korea	30.0	31.0	38.0	45.0
Japan	38.0	47.0	56.0	58.0
Urban transition in progress				
Malaysia	—	—	39.0	47.0
Philippines	26.0	30.0	33.0	39.0
Indonesia	25.0	27.0	32.0	36.0
Thailand	12.0	15.0	19.0	24.0
Incipient urban transition				
China	10.0	12.0	10.0	22.0
Vietnam	—	—	11.0	15.0

Sources: UNESCAP (1991: 2.31); Vietnam GSO (1990).

Kong, Japan, Korea, and Singapore) and an increase in the 50–60 age group. There are marked differences, however, on a gender and ethnic basis. In the next section we discuss gender differences in the urban labour forces of the region, international labour demand, and the persistence of informal sector employment.

The cultural meanings and practices associated with gender are of direct significance to the understanding of work and its implication on space and place. The study of gendered work patterns in and through space provides us with an intricate portrait of how and why certain places emerge, particularly in urban areas where a myriad of activities are performed in a relatively small geographical place. Cities are also characterized by complex divisions of labour, not the least of which is a gendered division, all of which are conditioned by and help reproduce certain spatial patterns (Massey 1984). It is now quite thoroughly established in Western feminist literature that men and women tend to inhabit cities quite differently with respect to perceptions of place, housing issues, leisure, and, of course, the world of work—both paid and unpaid (Keller 1981; Mackenzie 1989; Wekerle *et al.* 1980). The same can be argued for the situation found in many urban centres of the East Asian Pacific Rim which are the focuses of this paper.

Gender Differences in Evolving Work Patterns

This section will compare and contrast the work patterns of women versus men in the cities of the East Asian Pacific Rim. Conventional approaches to work either ignore gender differences when it comes to urban labour force characteristics or else discount the nature of unpaid and informal work, much of which is performed by women. Any discussions of women's work must take into account these spheres of labour as well as factors which contribute to the availability of female workers. These factors relate to the structure of the family and household and women's versus men's responsibilities in these social units. Whether the society in question is organized into extended families as opposed to nuclear families, or issues such as the availability of institutionalized day-care, condition the extent to which women are available for paid employment. Other related issues are the construction of masculinity and femininity in the society being studied and the ways in which women and men practise their day-to-day lives in light of these ideologies. These notions affect the appropriateness of potential activities and appropriate locations for 'work' to be performed (e.g. the home as opposed to the street). A number of these factors will be explored in relation to Asian men's and women's employment in this section.

A study of the female labour force participation rates (FLFPR) in selected East Asian Pacific Rim countries allows us to make interesting observations on cultural work patterns and their relationship to the work lives of women and adolescent girls.[9] Fig. 4.1 illustrates the 'm'-shaped curve for

[9] The authors have attempted to update the information provided by Gavin Jones (1984) which has been reproduced in Fig. 4.1. These data from the early 1970s pertain to the cities named or, in the case of Malaysia, cities with 70,000 or more inhabitants. More detailed studies include Heyzer (1986), Lim (1989), Murray (1991), Ong (1987), and Wolf (1992).

women's labour force participation which is typical of Western industrialized countries. Hong Kong, Japan, Korea, and Singapore exhibit FLFPR
which peak when women are in their twenties, decline in childbearing years,
and then increase somewhat in the late thirties following the rearing of pre-
school children.

By comparing our recent data and Jones's data of the early 1970s
(Fig. 4.1) we can identify some major changes. First, in the industrialized
countries of Japan, Hong Kong, Singapore, and Korea, the rate of female
adolescent labour force participation (age 15–19) has declined considerably
(Fig. 4.2). In the Philippines and Malaysia it remains moderate at nearly 30
per cent but in Thailand, the figure of nearly 70 per cent shocks the ob-
server. This may be related to young women's employment in factories,
domestic service, and perhaps as 'entertainers' (i.e. the commercial sex
industry in its various guises). Japanese and Korean women follow similar
patterns in the work-force over the course of their lives by peaking their
involvement in their early twenties, taking a five- to ten-year hiatus from
paid employment (presumably to have children and care for them in their

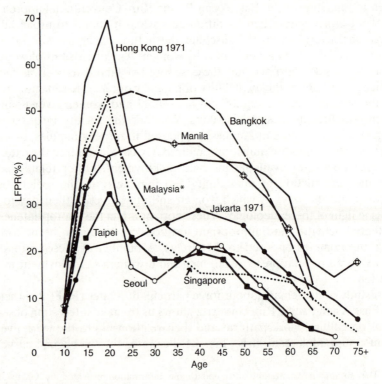

Source: Jones (1984: 29)

Fɪɢ. 4.1. Female labour force participation rates in selected Asian cities

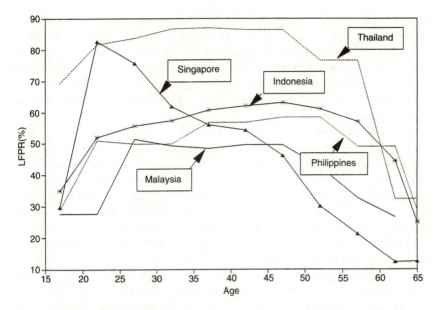

Source: ILO (1994).
FIG. 4.2. Female labour force participation rates in South-East Asia

infancy and pre-school years) and finally returning to the workplace by the late thirties. Thai women and Filipinas continue to follow what Jones described as the 'plateau'. Jones's discussions of the explanations of these curves, which point to the importance of 'culture', are still valid (Jones 1984).

In distinguishing between these three patterns of LFPR, Jones argued that the explanation for these curves was partly 'cultural' and partly related to 'levels of development'. In Singapore, Seoul, Hong Kong, and Taipei, as well as Malaysian cities, the Confucian-influenced patriarchal values manifest themselves in the primary involvement of women after marriage in looking after the husband and children. Another pattern typical of the Malay (Philippines, Indonesia, Malaysia) urban dwellers and Thailand is greater female participation in the working age group, reflecting a cultural practice where women's involvement in commerce is a long-standing tradition. Here, femininity is associated with earning income for the family. This was also related by Jones to lower levels of economic development where employment in the informal sectors of petty trade and handicrafts were more typical (Jones 1984).

Women are key agents in the urban development process in the East and South-East Asian economies. They play key roles in manufacturing, services, and the professions. Jose (1990) conveniently summarizes these trends based on recent labour force data. He notes that there has generally been an increase in FLFPR in all Asian countries except Thailand. Thai women have historically had the highest levels of work-force participation and Jose interprets the recent decline as due to withdrawal of child labour and older women following increased socio-economic development levels. This also may be related to women spending more time obtaining education and training and thereby delaying their entry into the work-force (Jose 1990: 61–2).

Women comprised between 30 and 40 per cent of all professional and technically skilled workers in all selected East Asian Pacific Rim countries except Korea which had less than 30 per cent and Japan which had 43 per cent (Jose 1990: 74). However, these female professionals are mainly concentrated in female type-cast occupations such as teaching and nursing. This reflects prevailing cultural attitudes that women are by nature 'nurturers' and inherently have the associated domestic skills required for these roles as a result. Administrative and managerial jobs are still dominated by men for the most part although the picture is slightly better in the Philippines and Singapore where women hold one-fifth of these jobs (Jose 1990: 75). Women play a very important role in the service sector where they dominate clerical and sales occupations in many of these countries (Jose 1990: 75–6). 'This is one occupational category where there has been a significant increase in the share of women workers in most countries' (ibid. 76). One area Jose neglects to address is the question of invisible home-based and 'informal-sector' labour where many women work either as 'contract out' workers or as micro-entrepreneurs. This topic, however, has been the focus of many case studies in Pacific Rim countries (cf. McGee 1971; McGee and Yeung 1977; Murray 1991; Sheridan and Salaff 1984; Tinker 1987). However, despite the predominance of women in services, women in blue-collar factory jobs have received far more attention (Heyzer 1986; Lim 1989; Ong 1987; Wolf 1992). This is ostensibly due to the overwhelming sex-typing of occupations in this sector where factory owners and managers specifically target young, single women to work the assembly lines because they are viewed as a docile, industrious work-force with 'nimble fingers'. Between 25 and 30 per cent of factory workers are female throughout Asia. Only in Thailand are the figures higher than 35 per cent (Jose 1990). Studies have shown that many of these jobs have been on terms grossly unfavourable to women (Jose 1990).

The general increase in female labour force participation rates in the urban areas of East Asia, while it reflects structural changes in the economies, is also related to systems of labour discipline that are reinforced by cultural values. Thus the systems of labour discipline and rewards are related to cultural understandings of gender roles. For example, women factory workers in Malaysia are given cosmetics as a reward for increased labour productivity. In many export zones, 'retrenchment' at times of reduced product demand is used by employers as a means of reducing the long-serving work-force who are earning higher wages. When they are re-employed they begin at the first level of employment. It is also assumed that women can be absorbed back into their homes at the time of retrenchment and that the employers are not responsible for the social costs of this dislocation.

Other mechanisms of social control in the labour process are the well-known example of company-based 'unions' that are mobilized to compete against other companies with rewards for higher productivity. In the broad-er context the general weakness of organized labour, whether for male or female workers, in East Asia is seen as a major attraction for both domestic and international capital. Thus culturally sanctioned systems of work disci-pline are central to the gender and other aspects of labour systems in East Asian urban areas.

The second aspect of urban labour force formation relates to the growing demand for specific labour skills in different parts of the region. In the industrialized cores as labour wages have risen and rural-sourced labour has become depleted there is a growing demand for unskilled labour to work in construction or other menial low-paid jobs. Thus, for instance, the 1980s in Tokyo have been characterized by a large inflow of foreign labour from China, the Philippines, Bangladesh, and Pakistan. These workers concen-trate in *Kichin* (cheap apartment houses in inner-city wards) creating ten-sion with the Japanese inhabitants (see Douglass 1993; Fowler 1993; Okuda 1990). Migration to the city thus affects housing structure, as it did in nineteenth-century Singapore and Hong Kong with the construction of shop-houses where workers were housed above commercial property. There is also a growing movement of female labour from countries such as the Philippines, to work as domestic servants in the major cities of the East Asian core. Sometimes, these arrangements involve recruitment into pros-titution activities (see Fadiar 1993).

Finally, despite the growth of the formal sector in most of these cities, there are still sizeable populations engaged in small-scale enterprise, in retailing manufacture and services often under conditions of self- or family employment. Some commentators who have attempted to explain the suc-cess of the East Asian economies attribute this to the persistence of this sector. For instance, in highly urbanized Japan, 88 per cent of all retailing units are still owned by families. In Seoul, old city markets persist and in some cities—Singapore for example—night markets have been encouraged as a promoter of tourism. It has also been suggested that the existence of the informal sector permits a continuation of 'subcontracting' and 'putting-out' which is beneficial to formal-sector enterprises. In the case of the socialist cities of China and Vietnam, the informal sector has been expanding rapidly as evidenced by the growth of 'free markets' in Beijing and Hanoi (Drummond 1993).

The Changing Nature of 'Work' in the Structural Transformation of East Asian Cities

In the preceding section we discussed the empirical features of the evolving labour forces of the East Asian Pacific Rim cities. In this part we want to relate the labour force changes to the 'nature of work', changing cultural values, social structure, and patterns of geographic mobility.

The results of the previous analyses indicate three important findings with respect to work. First, as the East Asian Pacific Rim countries experience economic growth and increasing urbanization they are undergoing a structural transformation in their roles from what have been described as 'producer' to 'consumption' cities. This reflects the growing incomes of the city population, the changing character of the labour market, and the shift to the provision of higher-order services. While this tendency is most marked in the cities of the industrialized core of East Asia it is also increasingly occurring in cities which until recently have been subject to less structural change such as Bangkok, Jakarta, and Shanghai.

Secondly, there is growing formalization of work with labour being carried out at regular times for wages, although there are still sizeable populations that fall into the 'casual labour' category and the lower productivity 'informal sector', which are becoming increasingly depleted as the 'formal sector' gains a greater control over the economy. This is especially true in countries such as China, Vietnam, and Indonesia.

Thirdly, there is a close relationship between the 'culture of consumption' developing in East Asian cities, labour force formation, and the various systems of work control. As 'white-collar' employment increases in Asian cities, so does the demand for 'consumer goods' such as clothes and accessories, which are an increasingly necessary part of the 'presentation' of the worker in the workplace. There are culturally sanctioned expectations of consumption which fuel these rapidly growing urban societies. As globalization proceeds, the 'culture of consumption' is increasingly influenced by 'globally' derived fashions and information which is delivered through increasingly obtrusive mass media which define popular urban culture.

Fourthly, there is ample evidence that these processes are creating in most cities a three-tiered social structure. In virtually every large city of the region, per capita incomes are four to five times that of the country at large. For instance, in Bangkok in 1990 almost 60 per cent of the households earned a monthly household income in excess of $US400 compared to only 14 per cent in the nation at large (see Hewison 1992). While these data mask a highly unequal distribution of income it also means that the East Asian cities are places where there is an increasing disposable income. The growth of consumerism in the cities of the region is evident in the urban landscape—particularly with reference to the construction

TABLE 4.6. Selected East Asian countries: ownership of selected consumer durables 1960–1985

	Motor vehicle (000s)		Telephones (per 1,000 inhab.)		Radio (per 1,000 inhab.)		Television (per 1,000 inhab.)	
	1960	1985	1960	1985	1960	1985	1960	1985
Urbanized countries								
Singapore	67	236	3.7	43.0	87	272	33.0	188
Hong Kong	32	161	3.7	44.0	55	518	2.0	228
Republic of Korea	13	557	0.4	22.0	41	451	0.3	175
Japan	419	27,844	5.9	55.0	133	713	64.0	563
Taiwan	6.3	830	0.9	24.0	42	236	191.0	244
Urban transition in progress								
Malaysia	97	1,392	1.0	8.0	—	410	6.0	113
Philippines	74	360	0.4	1.5	22	45	1.0	26
Indonesia	103	987	0.1	0.3	9	138	0.4	23
Thailand	41	688	0.2	1.9	6	205	2.0	27
Incipient urban transition								
China	—	—	0.4	0.7	11	—	0.03	7

Sources: UN (1987); Republic of China (1988).

of monstrous shopping plazas. East Asian 'yuppies', like their Western counterparts, are redefining urban life-styles and places. As this chapter will later demonstrate, local traditions, however, are incorporated into development trajectories in ways which challenge and modify 'Westernization'.

Statistical data to support this assertion are highly uneven for East Asian Pacific Rim cities but, even for comparatively poor countries like China, a high proportion of household monthly consumption (23 per cent) is spent on expenditures other than basic expenses for food, shelter, clothing, transportation, health, education, and fuel (see UNESCAP 1993: 4). Evidence of this increasing consumption is also available in the large increase in the ownership of automobiles and consumer durables (see Table 4.6). While these statistics are presented at a country level it is a well-known fact that ownership of these items is much higher in urban areas. These data reveal a dramatic increase in the ownership of consumer durables in the industrialized core of East Asia but countries such as Malaysia are now beginning to accelerate with the growth of the mass market. Another aspect of this growth of consumer expenditure is the increase in privately owned residences among the middle- and high-income groups which is leading to a rapid expansion of the periphery of East Asian cities with spreading suburbs.

All these developments are creating the conditions of mass consumption in the East Asian cities with dramatic consequences to the morphology and built environment of the urban centres. These patterns of consumption are, however, unequally shared by different socio-economic groups which is also a reflection of the occupational structure presented in the model of the urban labour force. Thus these divisions are reflected in the social structure of these cities.

Broadly defined, there appears to be a three-tiered stratum of social groups that have emerged in these cities. First, there is a small élite of the 'rich' consisting of large-scale capitalists, senior politicians, senior civil servants, and military officials who are the most affluent of the consumers. Depending on the country under discussion, a small group of upper-level managers are also included in this group. It is significant that this group is also emerging in the socialist countries as well as in capitalist societies. Secondly, there is a growing middle class made up of technocrats, white-collar salaried employees in government and capitalist enterprises, and small-scale capitalists; and finally a large low-income population are employed either as casual labour or in factory work and micro-enterprises of the informal sector.

While this threefold division oversimplifies a much more intricate social structure in which status and earlier historical social structures are still influential in the urban areas of East and Southeast Asia, it is a reflection of the changing structure of work and remuneration.

The changes are reflected in both the 'place' and space of the East Asian Pacific Rim cities. First, place of work and residence are now more clearly separated. The densely crowded cores of these agglomerations where low-income people commonly live at their place of work are being replaced by the ubiquitous skyscrapers of the modern inner city, commercial business, and escalating land prices. These inner-city populations are forced to move into low-income housing provided by the government. In Hong Kong, Singapore, and Seoul there is a shift to rental and low-income housing, or to the fringes of the cities as is the case in Bangkok, Taipei, and Kuala Lumpur. While some of the residents carry on their previous occupation, increasingly they and their families are absorbed into the formal sector. These changes separate the place of work and residence and since many of this group cannot afford privately owned motor cars they may utilize public transportation systems, which now in some of the countries include well-developed mass transit systems; but in others, such as Bangkok, Jakarta, and Manila, the systems of public transport are poorly developed and they rely upon various forms of paratransit such as 'illegal taxis' or 'jeepneys'.

For the middle class, these processes of economic and ecological change involve similar movements to the suburbs and longer journeys to work. This movement is, however, much more voluntary as the middle classes use their growing affluence to purchase automobiles and homes. In all the East and South-East Asian cities, despite the fact that different 'housing responses' have been developed such as apartments in high-rise buildings as is the case of Seoul and Singapore, suburbs are becoming ubiquitous.

The 'rich', while numerically quite small in these cities, are very observable. They cluster in wealthy housing areas such as Pondok Indah in Jakarta or Damansara Heights in Kuala Lumpur and they adopt a lavish life-style of frequent conspicuous consumption. For them, the journey to work is often made in air-conditioned and chauffeured cars. Those who can afford it are therefore sheltered from the ills of urbanization such as air and noise pollution which plague the cities of the East Asian Pacific Rim. The poor however are subject to the numerous health consequences of prolonged exposure to these hazards (Komin 1989).

As the workplace is increasingly becoming more formalized with regular hours there is more time for leisure and the taste for 'mass consumption' grows. As one writer in the Bangkok-based *Nation* suggested:

The winds of change in the Thai lifestyle are sweeping across this nation, with a speed that awes even the keenest observers. The vibrant economic growth has brought this country to a new level of affluence and with it a degree of Westernisation. Now it looks good and feels good to live and consume with style—the Western style. The era of mass consumption is upon us . . . The globalisation of consumerism transcends cultural differences and leaves the value of restraint as expounded by Buddhism of the past. (cited in Hewison 1992: 333)

This chapter argues that the process of socio-economic and cultural change in the cities of the East Asian Pacific Rim is far more complex than a term such as 'Westernization' can explain. First, influences of Japanese origin as well as Chinese origin are of key importance in addition to 'Western' investment, technology transfer, and way of life. More importantly, however, it appears as though East Asian Pacific Rim urbanites—most specifically élites—are borrowing ideas, styles, artefacts, and objects from foreign sources while at the same time making them their own by incorporating outside influences into the home culture in a unique way (Walker 1991). There is a growth of what appears to be Asian gentrification and post-modern design in the form of renovated districts and heritage architectural themes in housing, restaurants, and other spaces (see Yasmeen, forthcoming). The process of cultural change and *ex*change is therefore of a dialectical and dialogical nature rather than a question of a wholesale imposition or adoption of things 'foreign'.

This growth of mass consumption increases the number of non-related trips in the city. For the middle-class family, the weekend trip to the shopping mall modifies the seven-day work week. There is also time for leisure activities such as golf. By September 1991, one writer reported that 'there were 60 golf courses in operation throughout Korea with a further 118 under construction. Of these, 81 were within Seoul metropolitan area (28 presently in business, 53 under construction)' (Cotton and van Leest 1992: 363). While such leisure activities are normally limited to the wealthy élites, there is also a growth of middle-class-oriented theme parks of which the newest and most successful is Tokyo's Disneyland.

This growing affluence is also exhibited in increased mass tourism and business travel which has led to intense competition between East Asian Pacific Rim cities to attract tourists from both within the region and outside. This has led to a rapid expansion of employment in this sector.

These changes are in fact reflecting overall ecological and morphological changes in the East and South-East Asian urbanization pattern which is emphasizing the emergence of mega-urban regions centred on the largest cities. These cities are now encircled by new suburban developments, industrial estates, leisure park developments in which a growing proportion of the population of the cities are involved in the processes of material changes of the built environment. They rest upon the newly emerging 'transactional environment' of the city which is facilitating the segregation of places devoted to work, residence, and leisure but at the same time integrating the cities functionally.[10]

Conclusion

Today, while there are still great differences in the work patterns of the cities of the East Asian Pacific Rim, there appear to be common processes affecting these work patterns. In the West (particularly the USA) these processes have created urban areas in which segregation based on income, social status, and ethnicity has created cities of considerable social tension in which crime and violence abound. They have also led to major changes in labour processes involving increased open unemployment, changes in the location of work within cities, and an increase in part-time employment. Clearly, the governments of the East Asian Pacific Rim region fear that the rise of urbanization will lead to a similar experience. This is reinforced by the fact that it is impossible to segregate the residents of East and South-East Asian cities from the cultural impact of the West. This has been reinforced by fax machines, higher education, and jet travel.

Faced with this situation, Asian governments are seeking to find an 'Asian Way' which will prevent the worst aspects of the 'West' dominating their cities. Thus, they are turning to their cultural traditions as well as redefining the meaning of their cultures to stem the tide. Some believe that this can be found in the communitarian spirit drawn from Confucius but in

---

[10] See Ginsburg *et al.* (1991) for more discussion of the emergence of these Extended Metropolitan Regions. Also, Greenberg (1994).

a broader sense there is emphasis upon paternalism that expects the individual to be subservient to the state or the institution of which he or she is part. This involves respect for authority and community. It is often argued that government promotes these cultural values in order to create a more disciplined work-force. It is through such policies (most strikingly exhibited in Singapore) that cultural tradition is being drawn upon to maintain civil order in urban society just as it has to justify the discipline and social control of labour forces in the cities. The issues of integration versus segregation of women and men are also of key importance for understanding the spatialization of work patterns in urban places. In East Asia, women's labour force participation rates are on the increase which is a direct challenge to their place in the domestic private sphere. This, in turn, affects the nature of the public sphere and of public space which more and more women have access to.

The issue of the use of public space in East Asian cities is intimately related to the patterns of work that are evolving. As the work-force becomes more proletarianized and works more regularly there is more time for leisure activities and use of this public space becomes more individual. Likewise, the impact of the new middle classes on the city is overwhelming and reflected in the construction of shopping centres, theme parks, golf courses, suburban housing estates, and somewhat 'post-modern' renovated districts. Public space is still therefore used for collective spectacle in conformity with tradition. However, as the use of the space becomes more individual it also becomes the site for acting out social, political, and economic tensions such as strikes and political rallies, which in part are related to these new labour force formations. In such situations clearly the East Asian states would like to reassert the role of public space as 'respectful spaces' of obedience.

Whether appeal to the underlying 'cultural traditions' of East and South-East Asian societies will lead to a different 'culture of work' and management in the cities of the region is an open question, particularly if governments are unwilling or unable to deliver the social welfare that city populations expect. But perhaps a 'new urban tradition' is being forged in East and South-East Asian cities today. Thus, Lampard's insights on the role of the 'urban tradition' written almost thirty years ago may well be coming true in the East Asian Pacific Rim.

References

Agnew, John, Mercer, John, and Sopher, David (eds.) (1984). *The City in Cultural Context*. Boston: Allen & Unwin.

Appadurai, Arjun (1990). 'Disjuncture and Difference in the Global Cultural Economy', *Public Culture*, 2 (2): 1–24.

Armstrong, W., and McGee, T. G. (1985). *Theatres of Accumulation*. London: Methuen.

Arrighi, Giovanni, Ikeda, Satoshi, and Irwan, Alex (1993). 'The Rise of East Asia: One Miracle or Many?', in Ravi Arvind Palat (ed.), *Pacific Asia and the Future of the World-System*. Westport, Conn.: Greenwood Press.

Cotton, James, and van Leest, Hyung-a (1992). 'Korea: Dilemmas for the Golf Republic', *Pacific Review*, 5 (4): 360–9.

Deyo, Frederick (1989). 'Labour Systems, Production Structures and Export Manufacturing: The East Asian NICs', *Southeast Asian Journal of Social Science*, 17 (2): 8–24.

Dirlek, Arif (ed.) (1993). *What is in a R.?I.?M.? Critical Perspectives on the Pacific Region Idea*. Boulder, Colo.: Westview Press.

Douglass, Mike (1993). 'The "New" Tokyo Story: Restructuring Space and the Struggle for Place in a World City', in Kuniko Fujita and Richard Child Hill (eds.), *Japanese Cities in the Global Economy*. Philadelphia: Temple University Press.

Drummond, Lisa (1993). 'Women, the Household Economy, and the Informal Sector in Hanoi'. MA thesis, Department of Geography, University of British Columbia, Vancouver.

The Economist (1994). 'Asian Values', 331 (7865): 13–14.

Fadiar, Neferti Xina M. (1993). 'Sexual Economics in the Asia Pacific Community', in Arif Dirlek (ed.) (1993). *What is in a R.?I.?M.? Critical Perspectives on the Pacific Region Idea*. Boulder, Colo.: Westview Press.

Fowler, Edward (1993). 'Minorities in a "Homogeneous" State: The Case of Japan', in Arif Dirlek (ed.) (1993). *What is in a R.?I.?M.? Critical Perspectives on the Pacific Region Idea*. Boulder, Colo.: Westview Press.

Ginsburg, N. S., Koppel, B., and McGee, T. G. (eds.) (1991). *The Extended Metropolis: Settlement Transition in Asia*. Honolulu: University of Hawaii Press.

Greenberg, Charles (1994). 'Region-Based Urbanization in Bangkok's Extended Periphery'. Unpublished Ph.D. thesis, Department of Geography, University of British Columbia, Vancouver.

Haraway, Donna (1991). *Simians, Cyborgs and Women: The Reinvention of Nature*. London: Free Association.

Harvey, David (1989). *The Condition of Postmodernity: An Enquiry into the Origins of Cultural Change*. Oxford: Basil Blackwell.

Hewison, Kevin (1992). 'Thailand: On Becoming a NIC', *Pacific Review*, 5 (4): 328–37.

Heyzer, Noeleen (1986). *Working Women in Southeast Asia: Development, Subordination and Emancipation*. Milton Keynes: Open University Press.

Jones, Gavin W. (1984). *Women in the Urban and Industrial Workforce*. Honolulu: University of Hawaii Press.

Jose, A. V. (1990). 'Employment Opportunities for Women in Asian Countries: Progress and Prospects', in S. V. Sethuraman (ed.), *Employment Challenges for the 90s*. New Delhi: ILO-ARTEP.

Keller, Suzanne (ed.) (1981). *Building for Women*. Lexington, Mass.: Lexington Books (D.C. Heath & Co.).

King, Anthony (1990). *Urbanism, Colonialism, and the World Economy*. London: Routledge.

Komin, Suntaree (1989). *Social Dimensions of Industrialization in Thailand*. Bangkok: National Institute of Development Administration.

Lampard, Eric E. (1965). 'Historical Aspects of Urbanization', in Philip M. Hauser and Leo F. Schnore (eds.), *The Study of Urbanization*. New York: John Wiley & Sons Inc.

Lefebvre, Henri (1974). *La Production de l'espace*. Paris: Éditions Anthropos.

Lim, Linda Y. C. (1989). 'Women Industrial Workers: The Specificities of the Malaysian Case', in Jamilah Ariffin and Wendy Smith (eds.), *Malaysian Women and the Urban and Industrial Labor Force*. Singapore: Institute of Southeast Asian Studies.

McGee, T. G. (1967). *The Southeast Asian City*. London: G. Bell & Sons.

——(1971). *Urbanization in the Third World: Explorations in Search of a Theory*. London: G. Bell & Sons.

——(1985). 'Circuits and Networks of Capital: The Internationalization of the World Economy and National Urbanization', in D. Drakakis-Smith (ed.), *Urbanization in the Developing World*. London: Croom Helm.

——(1991). Presidential address: 'Eurocentrism in Geography: The Case of Asian Urbanization', *Canadian Geographer*, 35 (4): 332–44.

——(1995). 'Metrofitting the Emerging Mega-urban Regions of ASEAN: An Overview', in T. G. McGee and Ira Robinson (eds.), *Managing the Mega-urban Regions of ASEAN Countries*. Vancouver: University of British Columbia Press.

——and Yeung, Y. M. (1977). *Hawkers in Southeast Asian Cities*. Ottawa: International Development Research Centre.

McKay, Deirdre (1994). 'Engendering Ecologies: Interpretations from the Upland Philippines'. Paper presented at the Annual Meeting of the Canadian Asian Studies Association, Calgary, 10 June 1994.

Mackenzie, Suzanne (1989). 'Restructuring the Relations of Work and Life: Women as Environmental Actors, Feminism as Geographic Analysis', in Audrey Kobayashi and Suzanne Mackenzie (eds.), *Remaking Human Geography*. Boston: Unwin Hyman.

Massey, Doreen (1984). *Spatial Divisions of Labour*. London: Macmillan.

Morishima, Michio (1982). *Why has Japan 'Succeeded'?* New York: Cambridge University Press.

Murray, Alison (1991). *No Money, No Honey: A Study of Street Traders and Prostitutes in Jakarta*. Singapore: Oxford University Press.

Neville, Warwick (1993). 'The Dynamics of Population Ageing into the 21st Century: ASEAN and Selected Countries of Pacific Asia', in Yue-Man Yeung (ed.), *Pacific Asia and the 21st Century: Geographical and Developmental Perspectives*. Hong Kong: The Chinese University Prss.

Nonini, Donald M. (1993). 'On the outs of the Rim: An Ethnographic Grounding of the "Asia-Pacific" Imaginary', in Arif Dirlek (ed.), *What is in a R.?I.?M.? Critical Perspectives on the Pacific Region Idea*. Boulder, Colo.: Westview Press.

Okuda, Michihiro (1990). 'Changing the Face of Urban Life: As an Index of Megacity, Tokyo'. Unpublished paper presented to the Symposium on the Mega-City and the Future: Population Growth and Policy Responses, 22–5 October 1990, Tokyo. Sponsored by the Population Division, United Nations, and the United Nations University.

Ong, Aihwa (1987). *Spirits of Resistance and Capitalist Discipline: Factory Women in Malaysia.* Albany, NY: SUNY Press.

Palat, Ravi Arvind (ed.) (1993). *Pacific Asia and the Future of the World System.* Westport, Conn.: Greenwood Press.

Redding, S. Gordon (1993). *The Spirit of Chinese Capitalism.* New York: Walter de Gruyter.

Republic of China (1988). *Statistical Data Book.* Beijing.

Rimmer, Peter J. (1993). 'Transport and Communications in the Pacific Economic Zone during the Early 21st Century', in Yue-Man Yeung (ed.), *Pacific Asia in the 21st Century: Geographical and Developmental Perspectives.* Hong Kong: The Chinese University Press.

Said, Edward W. (1979). *Orientalism.* New York: Vintage Books.

Sheridan, Mary, and Salaff, Janet (1984). *Lives: Chinese Working Women.* Bloomington: Indiana University Press.

Sjoberg, Gideon (1968). 'The Preindustrial City', in Sylvia Fleis Fava (ed.), *Urbanism in World Perspective: A Reader.* New York: Thomas Y. Crowell Company.

Skinner, G. William (ed.) (1977). *The City in Late Imperial China.* Stanford, Calif.: Stanford University Press.

Tinker, Irene (1987). 'Streetfoods: Testing Assumptions about Informal Sector Activity by Women and Men', *Current Sociology*, 35 (3) (entire issue).

UN (United Nations) (1987). *Urban and Regional Development Statistics.* New York.

——(1992). *World Urbanization Prospects, 1990.* New York: United Nations.

UNESCAP (United Nations Economic and Social Commission for Asia and the Pacific) (1993). *State of Urbanization in Asia and the Pacific, 1993.* New York: United Nations.

Vietnam General Statistic Office (GSO) (1990). *Statistical Data of the Socialist Republic of Vietnam, 1976–1990.* Hanoi: Statistical Publishing House.

Vogel, Ezra F. (1991). *The Four Little Dragons: The Spread of Industrialization in East Asia.* Cambridge, Mass.: Harvard University Press.

Walker, Marilyn (1991). 'Thai Elites and the Construction of Socio-cultural Identity through Food Consumption'. Unpublished Ph.D. dissertation, Department of Social Anthropology. North York, Ontario: York University.

Wekerle, Gerda R., Peterson Rebecca, and Morley, David (eds.) (1980). *New Space for Women.* Boulder, Colo.: Westview Press.

Wheatley, Paul (1971). *The Pivot of the Four Quarters: A Preliminary Enquiry into the Origins and Character of the Ancient Chinese City.* Chicago: Aldive.

Williams, Raymond (1976). *Keywords: A Vocabulary of Culture and Society.* New York: Oxford University Press.

Williamson, J. G. (1988). 'Migration Selectivity, Urbanization and the Industrial Revolutions', *Population Development Review*, 14: 287–314.

Wolf, Diane (1992). *Factory Daughters: Gender, Household Dynamics, and Rural Industrialization in Java.* Berkeley and Los Angeles: University of California Press.

Wu, Yuan-Li, and Wu, Chun-hsi (1980). *Economic Development in Southeast Asia: The Chinese Dimension.* Stanford, Calif.: Hoover Institution Press.

Yasmeen, Gisèle (1992). 'Bangkok's Restaurant Sector: Gender, Employment and Consumption', *Journal of Social Research* (Chulalongkorn University, Bangkok), 15 (2): 69–81.

——(forthcoming). 'Bangkok's Foodscape: Gendered Access to Public Space in the City'. Ph.D. dissertation, Department of Geography, University of British Columbia, Vancouver.

Yoshihara, Kunio (1988). *The Rise of Ersatz Capitalism in Southeast Asia.* Singapore: Oxford University Press.

PART II

Dynamics of Urban Transformation: Case Studies of Six Cities

5

Studying Culture and the Built Environment

Kong Chong Ho and Won Bae Kim

New Orientations

From the 1980s, culture has enjoyed a renewed interest in urban studies (Leitner 1992; Jacobs 1993). However, within the anthropological tradition, this interest has a long history linked to the theories of cultural evolution in the nineteenth century. By examining all forms of human adaptation to the natural environment under the rubric 'built environment', early anthropological works viewed the built environment as an outcome of a group's culture. The particular forms themselves were seen to mirror the cultures that produced them (Lawrence and Low 1990). The more recent literature on culture and the built environment has produced shifts in the way culture is conceived and in the way culture is related to the built environment. Agnew, Mercer, and Sopher (1984), reviewing the collection of essays in *The City in the Cultural Context*, remarked that many of the essays used the more generalized notion of culture as a 'way of life' to culture as 'a realised signifying system'. However, with the influence of Geertz (see Jackson 1989) and the 'new cultural geographers' (Duncan 1990: 4), more recent works in anthropology and cultural geography, beginning from the late 1970s onwards, have stressed the second usage of culture as a signifying system.

This latter view of culture has shaped the approach to the built environment in the following ways.

1. If culture is a signifying system, and if the built environment is a cultural product, it is possible to study the built environment as a text to be read. In terms of anthropology, the shift is from a timeless, placeless language model of culture of earlier writers to a time-bound, place-specific, experience-rich, speech model of culture (Richardson 1989). Geertz's (1973) call for a 'thick description' ethnography is adhered to. In terms of Marxist urban studies, the move has been to bring culture out of the superstructure and to study it, along with politics and economics, as a basic determinant of material forms (Zukin 1988).

2. This new direction has resulted in a move away from the early focus on studying the relation(s) between man and environment to relations between people. The outcome (intended and unintended) of such interactions is etched in the built environment. The relationship between behaviour and the built environment is a reciprocal one, with the built environment affecting and being affected by social organization. Themes of domination and resistance, of boundaries and identity, permeate throughout the recent literature.

3. The new concern is also with an examination of ideas (e.g. salubrity (Rotenberg 1993); multiculturalism (Kobayashi 1993)), and the spatial amplification of such discourses in public and private spheres. This orientation can also be linked to the post-modern literature on the city, where the debate between modernity and post-modernity is related to an experience, a spirit of a particular socio-economic condition, or a reaction against the earlier periods and incorporates commentaries and research at several geographical scales, from the relation between economy and urban form, to architectural styles (see Cooke 1988 and Zukin 1988).

Studying Culture and the Built Environment

The new orientations promise a variety of insights from which to understand the role of culture and the built environment. From this variety, we extract a set of issues and features that collectively illustrate the various methods used to study culture and the built environment and allow the chapters in this part to be introduced.

Approach: symbolic and social production accounts

Stemming from the view of culture as a signifying system, the key concern of the symbolic approach is in viewing built forms as tangible evidence for describing and explaining the often intangible features of expressive cultural processes. Symbolic explanations often rest on demonstrating how the built environment corresponds to ideal conceptions of social, political, and religious life (Lawrence and Low 1990). The built environment is seen as a means of communication (Jacobs 1993).

The landscape is a central element in the cultural system, for as an ordered assemblage of objects, a text, it acts as a signifying system through which a social system is communicated, reproduced, experienced, and explored. The translation of cultural beliefs into the visible motifs of landscape exteriorizes that which was hitherto internal vision and thus helps to shape, control, and reinforce the internationalization of vision (Duncan 1990).

To argue that the built environment is socially produced is to be sensitive to the social, political, and economic forces that produce the built environment, and, conversely, the impact of the built environment on social action. The emphasis is on institutional forces and the changing historical and socio-cultural contexts within which the built form exists. Culture is usually referred to in terms of cultural plurality, and/or as a category, or in terms of ethnicity as a socially relevant category (Lawrence and Low 1990).

The primacy of economy and polity as factors accounting for the built environment, and the privileging of these over culture, is clear in King's (1988: 78) call for a political economy approach to the built environment: 'The general point, however, is clear: different forms of socio-economic organisation and political control, characteristics of phases of capitalism—or of socialist and other responses to it—have, with regional and cultural variation, produced forms of settlement and built environment and the cultural software to go with them.' Within such an approach, culture tends to be seen as a resource to be used in justifying action and symbolically reflecting dominance.

Whether the built environment is seen as socially produced, or a signifying system, is a question of emphasis, for the two approaches are not incompatible. The emphasis, therefore, rests on either an examination of the symbolic system or a focus on social relations. The choice is also in emphasizing the effect of the cultural system (as mapped on the built environment) on social organization, or in seeing the effect of social organization on the built environment.

The built environment can not only be seen as simple mirrors of culture. The built environment contains within it a variety of meanings. It was with the consciousness of this plurality that Agnew, Mercer, and Sopher (1984) preferred the term 'cultural context' rather than 'culture'. Increasingly, culture is viewed as dynamic and constantly being modified by existing social relations (Knox 1987). Studies within semiology are moving away from privileging the built environment and turning to an approach which takes account of social, political, and material contexts (Jacobs 1993). The symbolic meaning of urban landscapes has therefore been viewed as contested and fragile (Leitner 1992).

The set of chapters, to a varying extent, reflect this new orientation of incorporating symbolic and social production accounts of the built environment (see Table 5.1). The orientation is one of strategic emphasis. Of the six chapters, those on Beijing and Singapore concentrate primarily on political forces. The Beijing chapter examines how the notion of a political centre is being reconstructed by different regimes, while the Singapore chapter looks at how the changing relations between state and society are reflected in the contest over nature conservation, taking into account domestic and international forces shaping perceptions towards nature conservation.

TABLE 5.1. Overview of country chapters

Chapter element	Ch. 6: Seoul	Ch. 7: Beijing	Ch. 8: Tokyo	Ch. 9: Hanoi	Ch. 10: Hong Kong	Ch. 11: Singapore
Focus: Forces shaping built enviroment	Changing state, imperial, and cultural land-use policies	Political forces	Interplay of emperor system, state, and industrial capital	Impact of economic reforms (*Doimoi*); belief systems	British versus Chinese governments, value orientations of Hong Kong people	State versus civil society; global-local interface
Geography site	Sae'oon Sang'ga and Yoido projects	Tiananmen Square	Imperial Palace, Marunouchi district	36 Ancient Streets	Housing market, BOC building, airport project	Nature conservation areas
History periodization	Long view: history as clash of ideas	Long view: history as sedimentation, layering	Long view: history as sedimentation, layering	Long view: history as sedimentation, layering	Focused history: contemporary period	Focused history: illustrating state-society relation
Culture in relation to the built environment	Culture as resource: symbolic reflections of power	Culture as identity: power of place as a political arena	Culture as vision; Culture as identity	Culture as vision: influence of beliefs on the built environment	Culture as the value orientations of society	Culture as resource: the use of values, ideas to frame arguments

The Hanoi chapter concentrates on economic influences on the built environment, specifically examining the impact of economic reforms (*Doimoi*) on the cityscape. The Hong Kong and Tokyo chapters incorporate a broader analysis of forces, with Hong Kong looking at the conflicts between the British and Chinese governments, as well as an examination of how general value orientations shaped the landscape of Hong Kong. The Tokyo chapter, on the other hand, examines the interplay between the emperor system, the new state, and industrial capital in terms of their impact on the built environment at different historical periods.

Geographic scale: the site as a unit of analysis

Attempts to describe the built environment run into the inherent scale problem of finding an appropriate unit of analysis with which to represent the various forces. This problem is more urgent with the realization that culture, being fluid, is open to alternative interpretations, interpretations that are place bound. In this respect, we found Kuper's (1972) concept of site useful in solving the problem of dealing with an appropriate unit of analysis.

A site, as defined by Kuper (1972: 420), is 'a particular piece of social space, a place socially and ideologically demarcated and separated from other places'. By drawing on the analogy of a scene in drama, Kuper points out that sites are places where values are condensed, social relations are bounded. The contemporary literature also suggests that sites are places where activities are routinized and identities anchored. Therefore, from a methodological point of view, sites become important devices—windows or focal points—from which we can analytically represent the various forces at work.

The six case study sites include historically rich (or more accurately symbolically dense) areas represented by the Thirty-six Ancient Streets district or the old quarter in Hanoi, Beijing's Tiananmen Square, and the Imperial Palace and the Marunouchi district in Tokyo. The Singapore chapter focuses on what was until recently a marginal space, a variant of what McDonogh (1993) terms 'empty space': nature conservation areas. The Hong Kong chapter takes a more expansive view with the analysis including the Bank of China building and Chek Lap Kok airport. Along the same lines as the Hong Kong chapter, the Seoul chapter includes a variety of sites (monuments, buildings, the districts of Sae'oon Sang'ga and Yoido) as illustrations of changing state and cultural policies.

History and periodization

The fluidity of culture means that it is not only space bound but also time bound. A sense of history is essential in understanding the changes in social organization. And as society changes, the same landscape can take on different meanings (Leitner 1992). If the built environment is seen as a product of the *Zeitgeist*, or spirit of the age (Knox 1987), then a sense of history is also important in reflecting changes in this relationship. Thus, for these reasons, taking history into account is important methodologically for its ability to illustrate changing identities, meanings, social organization, new forces, and the relationships between these elements and their impacts on the built environment.

The six chapters illustrate different attempts at treating history. The Beijing, Tokyo, and Hanoi chapters take a long historical view and show how different periods resulted in the sedimentation of various forms and symbols. Zhu and Kwok condense the long 3,000-year history of Beijing and show how various elements of Beijing's built form reflect the notion of 'centre' or 'capital', and represent China's majesty, civilization, and authority. Trinh identifies four periods, the feudal, the colonial, the independence, and the post-socialist, in tracing the evolution of the old quarter at the centre of Hanoi. In a similar manner, Machimura tracks the imperial district of Tokyo through the Tokugawa period, the inter-war age of modernization, the post-Second World War period of industrialization, and the period up through the 1980s when Tokyo emerged as a global city.

Pai also takes a long view, but the historical process is represented as a clash of ideas and practices, not one of layering and sedimentation. Seoul's history is periodized into three phases: a long pre-modern period where the built environment conformed to the principles of *fengshui*, Confucian geographical thought, and the Korean conventions of hillside planning; the Japanese colonial period where imperialist and technical planning undermined the traditional symbolic order; and the contemporary post-war period which is influenced by a reaction against the colonial interventions of the previous period, and the push towards development.

The Singapore and Hong Kong chapter, in contrast, provides a selective or focused history tracing the development of state–society relations. Both the Singapore and Hong Kong chapters also attach greater emphasis to the contemporary period. The Singapore chapter traces changing relations between the state and society with regard to nature conservation in the last thirty years, while the Hong Kong chapter starts with the Joint Declaration of 1984 in detailing the relations between the British and Chinese governments.

Reading Culture in the Built Environment

Six city-specific chapters were cited in the previous section. Below, culture in the built environment is analysed under four themes.

Zeitgeist: *experience, cultural vision, and spirit of the period*

Of the six, the Tokyo chapter best illustrates the attempt to relate the built environment to the spirit of the age. Machimura has shown how, at different periods of time, the landscape of the Marounouchi district reflected the city as a showcase for novelties from abroad, as a powerhouse for innovation, a place of prosperity and contradictions in the post-war period, and the locus of post-modernity in the contemporary period. The Hanoi chapter also captures this essence for the contemporary period, as Trinh discusses not only the economic reform policies but also the change in social values, as the Vietnamese are caught in the transition from socialism to state-led capitalism.

Landscapes of domination and sites of resistance

Both the Beijing and Tokyo chapters contain rich accounts of how political authority is represented in the landscape of the capital cities. Machimura's use of the terms 'invisible' and 'visible' centre captures quite appropriately the influence of the emperor system and how this has incorporated—and at times dominated—the new political and economic elements at different periods in time.

More specific forms include Li's description of the Bank of China building in Hong Kong as representative of the tug of war between Britain and China, and Ho's description of the rejuvenated Singapore and Kallang rivers as monuments to showcase the government's achievements.

As pointed out in the first section, the landscapes of dominance have to be analysed alongside the sites of resistance (see Jackson 1985). In this regard, Machimura's analysis of Hibiya Park as a site for political activities and Ho's account of the community response over the government's plan to build a golf course on the Lower Peirce Nature Reserve provide the counterpoint to symbolic representations of authority.

Forms of engagement: language and ideas in contests

Several chapters also include conflicts over the built environment. Examples in the collection include the economic reform arguments versus conservation in Hanoi's old quarter, the contest between China and Britain over an airport project, and the issue of environment versus development debate in Singapore. Within these scenarios, culture becomes a resource in the sense that values, language, and ideas are appropriated by contesting groups to justify their positions and from which to build alternative conceptions. For example, the chronology of the environment versus development debate indicated how, as the state incorporated environmental goals, and as the public was influenced by global concerns for the environment, natural, undeveloped areas changed from being viewed as 'waste and swamp lands' to 'permanent national assets'.

General value orientations

The Hong Kong chapter shows how the combination of being ruled by a foreign government at a 'borrowed place' on 'borrowed time' resulted in materialism, utilitarianism, and political aloofness being the dominant values upheld by city residents. Li also shows how these features, materialism and utilitarianism, led to an intense speculative real estate environment, while political aloofness and apathy enabled a congested environment to be tolerated.

Trinh's analysis of Hanoi also indicated how the *Doimoi* open-door policy unleashed a new morality emphasizing consumerism and individualism, replacing the former socialist order of the collective, and which in turn threatens the destruction of Hanoi's old quarter.

These six country chapters, written by geographers, sociologists, and urban planners, attempt to link culture, social organization, and the built environment. Collectively, the set demonstrates an exciting variety of approaches to analysing the built environment, approaches that can be distinguished in terms of their choice of scales in time and place, as well as their treatment of culture, politics, and economics.

References

Agnew, John, Mercer, John, and Sopher, David (1984). Introduction and commentary, in John Agnew, John Mercer, and David Sopher (eds.), *The City in Cultural Context*. Boston: Allen & Unwin.

Cooke, Philip (1988). 'Modernity, Postmodernity and the City', *Theory, Culture and Society*, 5: 475–92.

Duncan, James S. (1990). *The City as Text: The Politics of Landscape Interpretation in the Kandyan Kingdom*. Cambridge: Cambridge University Press.

Geertz, Clifford (1973). *The Interpretation of Cultures*. New York: Basic Books.

Jackson, Peter (1985). 'Urban Ethnography', *Progress in Human Geography*, 10: 157–76.

——(1989). *Maps of Meaning*. London: Unwin Hyman.

Jacobs, Jane (1993). 'The City Unbound: Qualitative Approaches to the City', *Urban Studies*, 30 (4–5): 827–48.

King, Anthony (1988). 'Cultural Production and Reproduction', in David Canter, Martin Krampen, and David Stea (eds.), *Ethnoscapes: Environmental Perspectives*, i. Aldershot: Avebury.

Knox, Paul L. (1987). 'The Social Production of the Built Environment', *Progress in Human Geography*, 11: 354–78.

Kobayashi, Audrey (1993). 'Multiculturalism: Representing a Canadian Institution', in James Duncan and David Ley (eds.), *Place/Culture/Representation*. London: Routledge.

Kuper, Hilda (1972). 'The Language of Sites in the Politics of Space', *American Anthropologist*, 74: 411–25.

Lawrence, Denise L., and Low, Setha M. (1990). 'The Built Environment and Spatial Form', *Annual Review of Anthropology*, 19: 453–505.

Leitner, Helga (1992). 'Urban Geography: Responding to New Challenges', *Progress in Human Geography*, 16: 157–76.

McDonogh, Gary (1993). 'The Geography of Emptiness', in Robert Rotenberg and Gary McDonogh (eds.), *The Cultural Meaning of Urban Space*. Westport, Conn.: Bergin & Garvey.

Richardson, Miles (1989). 'Place and Culture: Two Disciplines, Two Concepts, Two Images of Christ, and a Single Goal', in John A. Agnew and James S. Duncan (eds.), *The Power of Place: Bringing Together Geographical and Sociological Imaginations*. Boston: Unwin Hyman.

Rotenberg, Robert (1993). 'On the Salubrity of Sites', in Robert Rotenberg and Gary McDonogh (eds.), *The Cultural Meaning of Urban Space*. Westport, Conn.: Bergin & Garvey.

Zukin, Sharon (1988). 'The Postmodern Debate over Urban Form', *Theory, Culture and Society*, 5: 431–46.

6

Modernism, Development, and the Transformation of Seoul: A Study of the Development of Sae'oon Sang'ga and Yoido

Hyungmin Pai

'Modernization of the Fatherland.' Though this translation of a key phrase in Korea's political jargon of the late 1960s conjures up unwarranted connotations, it none the less belies the aspirations and contradictions of the country's economic, political, and cultural policy of the period and beyond. Coinciding with the rapid industrialization and modernization of the 1960s, issues of a national cultural identity began to take on urgency. Almost always, however, culture has been reified into monuments and slogans that were somehow to bring order and identity to a rapidly changing society. Indeed a narrow, chauvinistic, and authoritarian discourse of culture was itself invented and cultivated as a compensation to the nation's economic and political upheavals. Focusing on the 1960s, this chapter looks at this peculiar configuration of tradition and modernity, culture and development, and its relation to the transformation of Seoul. It is a tentative search into the history of the mentalities and discourses that have so profoundly changed the most important city in the Korean peninsula.

Since the late 1960s, the speed and scale of urban development have only increased, and the rapid process of modernization has resulted in the destruction of almost all of the traditional urban fabric of Seoul. With the formation of private speculative capital, a boom in the overall economy, and the impetus of the 1988 Seoul Olympics, this fabric of Seoul has unfortunately proven to be all too vulnerable. The 1980s have witnessed the wholesale redevelopment of the city centre and the construction of super-block commercial and apartment complexes in the vast chequer-board land south of the Han river. The platitude 'unbridled development' takes on specific significance in the sense that development has been unhindered except in cases where sharp conflicts occur over economic interests. The more intangible forces of history and culture have proven to be minor in the expansion of the city. Admittedly, this kind of experience is not unique to Seoul. Unlike the capital cities of Europe, however, the devastation has been wholesale to the point where the palatial grounds of the Choson dynasty are virtually the only remnants of the ancient city. One may turn to the logic of capitalistic development for explanations. As the title of this book indicates, however, this chapter proposes to delve into the historical and cultural dimensions of this sudden and destructive process.

Underlying the theme of this study is the difficult problem of the complex relations between culture and the built form of the city. Rather than trying to mobilize the concept of a general, unspoken, and often implicit culture of nations, peoples, and social classes—culture as shared systems of beliefs, ideas, symbols, and practices—the chapter proposes first to examine the way culture (*munhwa* in Korean) was actually spoken of, represented, or was absent in certain projects, buildings and urban forms of Seoul. This is a study not of how the city has been used but how it is conceived and talked about. What has been said and built must first be established; one may then begin to examine what has been excluded.

The chapter is divided into three parts. The first section summarizes the formation and development of pre-industrial Seoul, outlining the special characteristics of its built environment. This is followed by an examination of the urban policies, transformations, and changing perceptions of the city during Japanese colonial rule. It will be argued that one must understand the peculiar and often tragic characteristics of the colonial experience to construe the nature of modernity in Korea. The third part then discusses two exemplary urban projects in Seoul during the late 1960s and early 1970s—the plans to develop Sae'oon Sang'ga and the Yoido area. The background to their development will be introduced and the related projects and discourses analysed. The two case studies are linked both historically and through the organizing concepts of the study. The goal of this chapter is to forge a connection between the urbanism of the 1960s and the particular configurations of modernism and the transformations of Seoul during the first half of the twentieth century. Through this study, the author hopes to shed light on the historical processes that resulted, on the one hand, in an urbanism that equated modernism with development, and, on the other, in development that was separated from any viable sense of culture. It is in this split between tradition and modernity, the absence of a dialectic between culture and development, that I believe the tragic nature of Seoul's urban transformation can be found.

Formation and Development of Pre-industrial Seoul

In 1394 Hanyang, which was one of numerous names attached to Seoul, became the capital of the newly founded Choson dynasty. Many influences can be seen in the planning of the city: the principles outlined in the *Kaogongji*, the tradition of *fengshui*, the emergent Confucian geographic thought of the literati, and the Korean conventions of hillside planning.[1] Under the requirements of the *Kaogongji*, the first decisions to be made were the sites of the main palace, Kyongbok Palace, and the ancestral temple of the royal family, Chong'myo. This was then followed by the careful placement of the shrine of the guardian deities of the state, namely the gods of the land and of the crops, to the right of the palace. These were the key places, along with the gates that surrounded the city, which formed the nodal points for the main thoroughfares. The common aspect of these influences, a point that I will return to later, is the importance of the natural landscape in the configuration of the city. The most representative would be the guidelines of *fengshui*, according to which the layout of the city was guided by the waterways and the geographical configuration of the surrounding mountains.[2]

The specific form of early Hanyang, however, is almost impossible to reconstruct. Most of the major monuments of the city as well as the residential areas were destroyed during the Japanese invasion in 1592. In addition, no census, registration, or maps have survived the devastation of the war. In fact, the Choson dynasty suffered from continuous invasions and internal strife, culminating in the invasion of the Manchurians in 1636. In this tragic period much of the original form of Hanyang was obliterated, and it took more than a century to reconstruct and rehabilitate the city. By the mid-eighteenth century Seoul had entered a period of prosperity, a time of both stability and change. It experienced an increase in population, the development of commerce, expansion of its urban area, and transformations of social structure, a phenomenon that had already begun by the end of the seventeenth century (T. J. Lee 1994).

Though specific data concerning pre-industrial Seoul are difficult to ascertain, the general mode of its social structure and urban form can be described. For example, recent studies of seventeenth-century censuses of the northern part of Seoul have estimated that servant slaves constituted about 75 per cent of the total population of the city, of which a large portion were live-in servants of the ruling class. Royalty and the ruling literati, constituting less than 10 per cent of the population, resided in districts adjacent to the palatial areas north of Chong'no, the central east–west thoroughfare within the walled city (Cho 1992). Low-level clerks,

---

[1] There is much debate on the specific historical influences on Seoul, particularly the relative importance of *fengshui*. For example Lee Tae Jin stresses the importance of new and rationalistic Confucian thinking as the primary force in the planning of Hanyang. He argues that the influence of *fengshui* has been over-emphasized, the result of a mythical reconstruction of the founding of Hanyang during what he calls 'the crisis period of 17th century Chosun' (Lee 1994). The importance of hillside planning principles can be found in Lee Sang Koo (1994).

[2] For an extensive account of the construction of the new capital, see Seoul Metropolitan Government (1977: i. 195–342).

peasants, merchants, and artisans constituted a plebeian class, literally called the 'middle-people' (*Joong'in*). True to their social designation, this class worked and lived in the middle part of the city, south of Chong'no. In spite of their population, the servant and plebeian class wielded little civic and economic power. Hanyang may thus be characterized as a typical 'orthogenetic' city—a political machine used by a ruling class whose economic basis of power lay in the agricultural production of the countryside.

In terms of its urban form, despite the affiliation to the classical Chinese city, it is difficult to see Seoul as a grid city in the fashion of Changan and Rakan. The street patterns did not form regularized grids, nor was there a central axis that symbolically ordered the social and physical environment. The most visible man-made artefact of the city was the gates of the encircling wall, the most prominent being Kwanghwamoon, the ritual gate of Kyongbok Palace. The north–south thoroughfare between Kwanghwamoon and Chong'no functioned as the administrative centre of Hanyang. Even in this most symbolic of places, it was less the control of artefacts—exemplified in the West in places such as Versailles, where man-made monuments exerted dominance over nature—than the totality

FIG. 6.1. Aerial view of Chong'no at the turn of the century when the traditional morphological structure of the area was still intact

of the natural, social, and ideological environment that shaped the nature of the place.

The morphological configuration of pre-modern Seoul may thus be characterized as large patches of urban areas unevenly developed and physically demarcated by extended one-storey structures (*haeng'nang*) that often housed commercial functions. Within the large block areas delineated by the main thoroughfares, the morphological pattern of residential areas consisted of loosely organized courtyard houses of various sizes accessed by narrow culs-de-sac (T. J. Lee 1994) (Fig. 6.1). Hanyang shows many of the characteristics of the classical East Asian urban tradition, i.e. the city as a social, political, and religious order.

This sense of the city as an embodiment of social hierarchies can be seen physically in the system of land regulation and residential differentiation according to social class. Initially, the laws of the *Kyungkukdaejun* provided guidelines as to the distribution of land and the size of houses. During the Choson dynasty, private ownership of land was in principle not acknowledged within the city limits of Hanyang. Residents were merely awarded the right to use and build on a designated piece of land. The size of one's land and residence was naturally regulated according to the social class of its owner. For example, first-level cabinet members of the ruling bureaucracy were awarded land no larger than 2,112.45 metres. Though there were restrictions on the height of the building and architectural elements and ornamentation, the primary object of regulation was the horizontal expansion of the house. The size of the houses of royalty and high-level officials was described in terms such as 'ninety-nine *kans* (bays)', 'nine (horizontal) level palaces', 'twelve gates', 'six inner courts'. From the beginning of the Choson dynasty, however, the city was in constant shortage of land, and the only way of providing land for the increasing populace was to divide and subtract from parcels that had already been distributed. Not surprisingly, powerful families often abused this fuzzy concept of land ownership and illegally expanded the borders of their residences by buying out their neighbours. This incremental pattern of expansion and division was further complicated with the breakdown of the strict class system after the Japanese and Manchurian invasions. Massive resettlement coincided with this social dissolution, resulting in further additions to and divisions of the physical environment.

The structure of the traditional city, from the scale of the individual house to the whole urban configuration, can thus be characterized as being 'deep' in both social and physical terms. First of all, it was a deep city in the sense that it had to sustain the various levels of social differentiation. Physically, Hanyang was a city of many horizontal layers, an environment formed through incremental additions, divisions, and subtractions. It sustained the strict class divisions of early *yangban* society as well as the more disorganized structures after the eighteenth century. Here we have a specific historical example of a city of multiple landscapes, a concept discussed by Mike Douglass in Chapter 3 above. An important point to be recognized is that artificial structures were not the main tools in ordering and symbolizing this multiple landscape. In a study of the traditional landscape structure of Seoul, Kim Han Bai notes that 'historical elements play a more important role as communication media (in cognitive experience) than as visual objects, because the invisible meanings they transfer are stronger than their visible characteristics' (Kim 1993: 337). He thus characterizes this type of landscape as organic, multiple, and internal (Kim 1993). The symbolism of power depended less on what one saw than on what was unseen, what was hidden. But, as will be shown, in the course of Seoul's turbulent history, there were fundamental shifts in the nature of its built environment and the manner in which it was perceived.

The City during Japanese Colonial Rule

It is the twentieth century that a new symbolic and morphological order was introduced to Seoul, and with it emerged a new consciousness of the city. Naturally, the transformations of the city were inextricably linked with the larger issue of modernizing the country and its people. The experience of Korea, and Seoul, however, is characterized by the fact that much of the process of modernization overlaps with the experience of colonial rule by the Japanese during the first half of this century. In the twenty-year span between 1920 and 1942, the urban population of the country increased from 3.4 per cent to 14.4 per cent (Kim 1971). Though these numbers do not compare in absolute terms with European or Japanese rates of urbanization, the impact of urbanization was fundamental and irrevocable. Numerous scholars have stressed the fact that Korea had been going through its own dynamics of change when the intervention of foreign imperial powers distorted this historical process. Urbanization, and the responses to its attendant problems, have also been interpreted in this way. With the growth of Hanyang since the eighteenth century, it had already begun to experience typically modern urban problems. It is evident that the development of commerce and the physical expansion of the city began to strain the ability of the deep city to sustain such new developments. By the late nineteenth century, the need for rational methods of dealing with these problems, particularly the task of clearing main thoroughfares that had been illegally occupied by squatters and merchants, was discussed by reformists such as Park Young-Hyo and Kim Ock-Kyoon. These efforts at reform, however, were fundamentally limited because the city could not deal with the basic issue of land ownership (Kim 1991). Subsequently, the first systematic transformation of the city was brought about by the

Japanese colonial government between the first years of this century and
their exodus from Korea after their defeat in the Second World War. It is
thus important to consider the implications of the modernization of Seoul in
the hands of an alien and malicious power.

The initial stages of colonial policy towards Seoul, or Kyungsung as it was
now called, were first of all to use it as a tool for the efficient exploitation of
the country and for further military expansions into China and Russia, and,
secondly, to put in place a new symbolic order that would match its imperial
power. For the Japanese to exploit fully the land resources of Korea it had
to be reorganized in a rational and systematic manner. It is not surprising
that one of the first priorities of the colonial government was to conduct an
extensive land survey of the country, thereby establishing clear boundaries
of ownership that would facilitate the expropriation and exploitation of
agricultural as well as urban land. Rational techniques of land subdivision—
originally adopted by the Japanese from the German practice of *Umlegung*,
the repartitioning of lots for city expansion and building—were adopted in
redeveloping and expanding urban areas (Seoul Metropolitan Government
1984). With this practice, the traditional incremental pattern of urban development was replaced by a system in which clear patterns of private land
ownership were drawn before building could occur. For example, a 1928
city planning report contains a redevelopment plan, to be implemented by
repartitioning techniques, for the city centre of Seoul. After designating the
five districts to be redeveloped, it noted that in spite of its central location
and commercial functions, and in spite of its linkage with major traffic lines,

because of the uneven shape of the blocks, the irregular curves, length and width of
the roads, the many cul-de-sacs, the area [did] not have a proper urban system and
in terms of hygiene, there [was] lack of facilities for drainage. Therefore, when rain
[fell] there [was] often flooding and in terms of traffic, safety and hygiene, there
[were] many degraded residential areas. Consequently, it [was] necessary to quickly
implement land repartitioning. (Kyungsungboo 1928: 3)

Traffic, safety, and hygiene—here we see one side of the urban policy of the
colonial government: a strategy of rationalizing the city, constructed in
the typical discourse of a positivist urbanism, devoid of aesthetic or
cultural language.

The other side of the Japanese policy, as I have already mentioned, was
to reorganize the city centre of Seoul according to a new symbolic order
(Rhee 1990). Though never fully realized, the Japanese plan proposed to
transform the main roads of the city into a grid system, overlap it with
baroque diagonals, and place large symbolic buildings in the key nodal
points (Fig. 6.2). The key monument in this scheme was the headquarters of
the colonial government. In 1924, Kwanghwamoon was moved to the east
of the Kyongbok Palace, and a domed, proto-Renaissance building was
erected in front of the main pavilion of the palace. The idea was that the
new building would break the original, spiritual axis of the city, so carefully

laid out during its foundation. The headquarters building was connected with City Hall, built in a simple and heavy classical style, and then linked to Namdaemoon, and further south-west to the Seoul railway station (see Fig. 6.3).

Often the two goals would be achieved at the same time, as in the case of constructing the necessary infrastructure that linked Seoul with the newly established industrial areas of the country. Though Seoul had already expanded well out of the inner wall during the nineteenth century, the railway linked with inner-city electric tram lines literally tore down the walls of the city, and with it a symbolism that had lasted for many centuries. The ideological and technical manœuvres of colonial urbanism thus sought to undermine the traditional symbolic order of Seoul and, in the process, introduced a new set of urban principles. A new symbolic order of perspectival vision and a scale controlled by monumental artefacts had been

FIG. 6.2. Urban renewal plan of Kyungsung, 1928

Note: The numbers denote the new roads to be built or expanded. 1, 2, and 3 constituted the new axis from the colonial headquarters to Namdaemoon, while 36 and 37 were the diagonal roads that were unrealized.

FIG. 6.3. View of Kwanghwamoon area before and after the construction of the colonial headquarters

established. By the 1930s, the most visible monuments in Seoul had become the steep towers of Myungdong cathedral, the massive presence of the colonial government building, and the mecca of a new culture of commercialism: Mitsukoshi department store. These 'Western-style' buildings and the urban environment they created departed from the deep structure of Seoul's traditional urbanism. The transformations they brought about were sudden and violent. Rather than being additions to the existing fabric, they acted as the centre of a new order. They dominated the landscape, providing explicit, in the physical and visual sense, but incomprehensible environments.

These new technologies, shapes, and symbols were to many embodiments of modernity itself. One well-documented instance is the image of the train breaking through the old gates of the city, as represented by Choi Nam Sun's 1908 poem 'Ode to Kyungboo Railway':

> With the roaring burst of the whistle
> and its back turned against Namdeamun
> it departs like the swift wind
> and even a bird with wings may not catch it.

The city thus emerges as the locus of a modernity of technical dominance over the old world problems of medieval Korea, of modernization and democratization, something that many of its reformist intellectuals had longed for. The 'iron horses' that traversed the urban landscape and the factory chimneys spewing smoke into Seoul's air were regarded with awe and admiration.

The second aspect of urban transformation during the colonial era brought about a different face of modernity. It is a period that coincides with a shift in the Japanese colonial strategy towards a softer 'cultural policy' during the 1920s—a policy of placation and coercion, always underlined, however, by the threat of violent oppression. By this time, a large Japanese residential area had emerged in the southern part of the city, in what is now called the Yongsan area. Centred on the Mitsukoshi department store, a modern commercial, banking, and entertainment area was formed between the Namdaemoon and Yongsan area. It became the centre of a new consumerist and cosmopolitan culture. During this time, the first strains of an 'anti-urbanism' comparable to that of nineteenth-century Europe and America began to emerge in popular and artistic discourse. It seemed to many that colonialism, urbanization, capitalism, and modernization were in fact one and the same thing. It was the force behind the decadent and alien urban landscape, bringing with it prostitution, commercialism, and irrelevant cultures. It would seem that during the late 1920s and 1930s the notion of an urban culture, i.e. the consciousness and practice of a way of life peculiar to the modern city, was raised and made explicit. Nowhere is this more evident than in the proliferation of literature concerned with themes of the city and its contrast to rural life. Whereas much of the artistic energy of the previous decades had been focused on themes of ideology and nationalism, the city emerged as the specific space of contradictions. As Lee Jae Sun notes, 'the general way of life of the city was portrayed not only in terms of poverty, crime, pleasure and prostitution, the biological friction and psychological tensions of human relations, alienation and individual schizophrenia, but the atmosphere of the city was presented as a microcosm of colonial life, with utterly no space for freedom' (Lee 1979: 321).

During the first half of this century, Seoul thus witnessed an unhappy marriage between urbanism and culture. The formation process of the built environment was fundamentally altered and the basic condition of capit-alistic urban development, that is the delineation of private ownership of land, had been installed. However, the potential capital necessary for urban development was nearly all in the hands of the Japanese. Furthermore, colonial policies were geared towards domination and a symbolic display of power. On the one hand, the city was the site of modernization, but one that was led by alien subjects. Toward the end of Choi's poem, it is acknow-ledged that the modernization is not that of the Korean people.

> This train I ride from morning to night
> I ride as if it were mine but in truth it is that of the other
> when shall we grow strong
> so that we may run it with our own bare arms?[3]

This was then the modernist dilemma of the colonial period. To be on the side of modernization was to be a Japanese sympathizer and a traitor, as with the tragic cases of Choi and Lee Kwang Soo, to the Korean race.

On the other hand, the city is the 'iron cage' of all that is bad, in the most essential aspects of life, in the experience of colonial domination. The new patterns of the city did represent the dismantling of the old regime of class society. The new élite class of officials, industrialists, and landowners were, however, mostly Japanese, and, furthermore, it is estimated that more than 35 per cent of the residents of Seoul were living in abject poverty.[4] The distortions of colonialism prevented the growth of a public sphere, grounded on the struggles inherent in the formation of new structures of modern society.

For most Koreans, the possibility of intervening in the city as subjects and objects, i.e. the possibility of a modern urbanism, was excluded at the outset of the twentieth century. The new urban environment was conceived almost solely as an alien other. As I have already mentioned, to accept it was to be a traitor. One could certainly reject it, but only in terms of who built it, who occupied it, and why it was built. Absent in colonial modernity was an immanent critique, a continuous criticism of the self linked with a trans-forming practice. This was a very specific form of alienation between the residents of the city and its built environment. The parameters of modern culture and its representations were rigidly confined by the sym-bolic, political, and economic impoverishment of colonialism. For many

---

[3] I am indebted to Jung (1991), for calling attention to Choi's poem and its implications for understanding an important thread of modernist thinking in Korea.

[4] For a brief account of the social structure of Korea during the Japanese colonial period see Kim (1984).

years to come, the legacy of colonial modernity would continue to haunt the country, forcing its soldiers, politicians, and intellectuals merely to react to this history. In terms of its urban consequences, this legacy is important as much in terms of its institutional and ideological formations as the actual built environment, transformed during a period of more than four decades.

Sae'oon Sang'ga and the Yoido Area

After the end of the war in the Pacific, the Cold War split the country into two. The sudden departure of the Japanese, the uprooting of people from their homes and land, and the vast influx of poverty-stricken people brought about a chaotic situation in Seoul. The devastation of the Korean War left the city in further disarray for more than a decade. Though there was some semblance of order during the late 1950s, the fall of Rhee Seung-man and Jang Myun's government, the former to a popular uprising in April 1960 and the latter to Park Chung-Hee's military coup the following year, prevented any systematic intervention in the city. At the same time, Seoul began to become an unwieldy and expanding city. In 1963, its population reached 3 million, and by 1970 it would increase in dramatic fashion to 5 million. The next stage of intervention in Seoul began during the mid-1960s, with the developmental policies of Park's regime. In fact, the year 1966 clearly marks the beginning of radical changes in Seoul. The basic tenets of the Second Five-Year Economic Plan, emphasizing industrialization and development, were announced in the great drive toward the 'modernization of the Fatherland'.[5] Seoul implemented its modernization policy under the newly appointed 'bulldozer' mayor Kim Hyun-Ok. His tenure between April 1966 and April 1970 saw the planning and implementation of major urban projects such as the initial land subdivisions of the Kangnam area, CBD redevelopment, and the building of major throughways in Seoul, as well as the development of Sae'oon Sang'ga and Yoido, the two case studies of this chapter.[6]

At the same time, the nationalist cultural policy of the military regime began to take shape. In the decade between 1966 and 1975, no less than ten major cultural facilities—the National Museum, Buyo Museum, Kyungjoo Museum, Pusan Municipal Museum, Sejong Cultural Centre, to name just a

[5] Though reiterated throughout the 1960s and early 1970s by President Park, the most basic ideas of the Second Five-Year Economic Plan, in perhaps its most rhetorical form, can be found in his New Year's Message to the National Assembly in 1966. Reprinted in the *Developing Forum*, 1 April 1966, of the Korea Planners' Association.

[6] Like many of the high-ranking officials of the Park regime, Kim Hyun-Ok came from a military background, and, receiving the trust of the President, was perhaps the most powerful mayor in the history of the city. Before coming to Seoul, he had been the mayor of Pusan, where he earned the nickname of 'bulldozer'. His downfall came, appropriately, when an apartment building that was just a few months old collapsed, killing 33 residents and injuring 40.

few—were either planned or built and, moreover, were to be designed to
emulate traditional Korean architecture. In 1968, the Ministry of Culture
and Information was established in order to set strict restrictions on art and
everyday life. The government also initiated various cultural projects, such
as the movement to use only *Hangeul*, the Korean alphabet, and eliminate
all use of Chinese characters, and English and Japanese words. To take an
architectural example of this nationalist policy, one may look at the competition for the new National Museum. Announced in January 1966, the
competition programme specified that the new building take elements
directly from existing ancient monuments. Another example, perhaps the
most important of many counter-reactions to the colonial transformation of
Seoul, was the return of Kwanghwamoon to the centre of the Kyongbok
Palace axis, right in front of the former colonial government building. A
statue of Lee Soon Shin, an admiral and hero of the Japanese invasion of
1592, was erected at the centre of the axis by the Committee for Erecting
Sculptures of Patriotic Forefathers, one of many public sculptures that this
organization commissioned during the late 1960s and early 1970s. Culture,
as it was formulated in the official discourse of the 1960s and beyond, was
national, ethnic, and traditional.

It should be stressed that this nationalist cultural policy emerged on the
heels of the normalization of relations between Korea and Japan in 1965.
Normalizing relations with Japan was extremely unpopular but crucial for
the Park regime in implementing its policy of bringing in foreign capital for
the development of Korea. It is thus no coincidence that many of the large
projects of the late 1960s, funded by loans from Japan, are named after war
heroes and independence movements in the country's struggle with Japan,
such as the Sam'il Expressway, named after the independence movement of
1919. On the one hand, this may be viewed as a form of cultural compensation and ideological manipulation. On the other hand, we have a form of
wish-fulfilment, one that lingers on from the dilemma that faced proto-
modernist intellectuals during Japanese rule such as Choi Nam Sun and Lee
Kwang Soo. For the modernist intellectuals of the 1960s, it seemed that
finally the means of modernization had come into the hands of an independent Korean nation.

It was in this milieu of the late 1960s that the Sae'oon Sang'ga and
Yoido projects emerged. Sae'oon Sang'ga was the first official redevelopment project of Seoul. It is comprised of four rectangular blocks of
mixed-use facilities, linked by pedestrian overbridges and decks, that
traverse the four main east–west thoroughfares of Seoul (Fig. 6.4). The
origins of its site can be traced back to 1944, when the Japanese designated
a 50-metre-wide and 1.2-kilometre-long area as a fire zone and ordered the
clearance of all buildings.[7] When the war in the Pacific ended the following

---

[7] During the last stages of the Second World War, fire zones were created by the Japanese

FIG. 6.4. Sae'oon Sang'ga during construction

year, about 80 per cent of the area had already been cleared. With the abrupt departure of the colonial government and the ensuing confusion, the area began to be occupied by squatters and refugees from North Korea. Furthermore, influenced by the upper-class brothels located north of Chong'no, the area became a red light district, a sore in the eyes of city administrators.

In 1951, hard as it may seem to believe, a plan was devised to demolish Chong'myo, the ancestral temple of the royal family, and erect a new National Assembly building on the site. Under this plan, the 50-metre-wide road would be transformed into the symbolic axis of the National Assembly parallel to the Kwanghwamoon–Namdaemoon axis. This idea was fortunately overturned when the Department of Cultural Artefacts vehemently objected, and, perhaps more importantly, when President Rhee Seung-man moved the site of the Assembly to an area in Namsan where the Japanese had built a Shinto temple. Subsequently the residents of the area argued that, if properly planned and built with modern construction materials, it would no longer be a fire hazard, and that they should be able legally to acquire ownership of the land. In fact, towards the end of the Rhee government, the ownership of about half of the land had been transferred to the private sector. This process was interrupted when once again, this time due to the legislative branch, the Chong'myo site became a candidate

colonial government in preparation for the expected bombing raids by Allied forces. For a detailed account of the background to this history see Sohn (1990: 311–38).

for the new National Assembly. The confusion continued on until the aforementioned Kim Hyun-Ok was appointed mayor of Seoul. Though debate and squabbling between officials of the city and central government, planners, architects, and residents over how to develop the area continued, the new municipal government pushed through plans that resulted in the present structure which was completed in 1967 (Korean Administrative Research Institute 1980).

At about the same time that Sae'oon Sang'ga was completed, plans to develop a large island of 2.6 kilometres square on the Han river, used as an airport at the time, into the 'Manhattan of Korea' was under way. It was part of what was called the 'Supreme Master Plan' of developing Seoul into a multinuclei mega-city. The original plan for Yoido was devised by the young but prominent architect Kim Swoo Geun (1931–1984), who had also designed Sae'oon Sang'ga.[8] The plan was organized by linear mega-structures that expanded along an east–west axis linking the 'National Assembly district' and the 'City Hall district'. Commercial and office functions were concentrated along this axis while housing was located next to this central spine. This axis was traversed by a north–south bridge crossing the Han river and a civic plaza was placed at the crossing point (Fig. 6.5). The plan owes much to Kenzo Tange's proposal for Tokyo 1960, which was also notable for the mega-structures that traverse Tokyo Bay. The common key to both proposals was the creation of a multilevel transportation axis, separating vehicular and pedestrian traffic, around which the whole plan was organized. Like Tange's plan, which aimed at shifting the 'radial and centripetal system' of Tokyo to a 'system of linear development', Yoido was to be the crucial link in the plan to transform Seoul into a linear metropolis that would extend west to the port city of Inchon. Like Tange, who envisioned that his plans would 'find the means of bringing the city structure, the transportation system, and urban architecture into organic unity', the planners of Yoido saw their work as an attempt to 'transform the fundamental form-structure of the city' (Tange 1993: 330–2; Korean Engineering Consulting Corporation 1969: 15).

The utopian plan by Kim, however, fell through, and a conventional plan that followed existing subdivision practices was adopted in the early 1970s. Once again, the National Assembly building came into play as the controversial issue was settled, at least in the matter of its site, when it was built in the north-east end of this newly developed area. In place of the futuristic

[8] Kim received his architectural training at Seoul National University, Tokyo Art University, and the University of Tokyo. While still a graduate student, he began his meteoric rise to fame. He and a group of students studying in Japan were awarded first prize in the competition for the National Assembly sponsored by the Jang Myun government. His rise continued when the Park government provided him with further major projects that monumentalized the power and aspirations of the new military regime.

Fɪɢ. 6.5. Model of the Yoido plan, 1969

civic-transportation axis, a large tract of asphalt measuring 0.25 kilometres by 1.5 kilometres came to traverse the island. Named after Park Chung-Hee's *coup d'état* of 1961, the 5.16 Square (16 May Square) has been used mostly for mass religious gatherings and military parades. Yoido also became the first area in which super-block apartment complexes, at present the dominant landscape of large tracts of land south of the Han river, were developed in full scale. It is currently an assemblage of super-block apartment complexes with their own schools, large and medium office buildings, the tallest skyscraper in Korea, the National Assembly, headquarters of television networks, one of the largest churches in the country, and the 5.16 Square, now renamed the Yoido Square.

In creating the plans for Sae'oon Sang'ga and Yoido, Kim Swoo Geun was acting as the vice-CEO of a larger semi-public organization, the Korean Engineering Consulting Corporation (KECC). From 1965, the KECC functioned as the primary engineering and development consulting firm of Park's military government.[9] It was to become the planning agent of its new

---

[9] In 1963 the KECC originated as the International Industry Technological Group, a small private firm located in Kim Swoo Geun's atelier in Seoul. The following year, it was renamed the Korea Pacific Consultants, and in August of 1966 acquired the KECC label and Kim's architectural office. See Jung (1996: 79–80) for a brief history of the KECC and its relation with Kim. The author of this book argues that Kim Jong-Pil, President Park's right hand man during the 1961 *coup d'état* and the chairman of the ruling Republican Party, was the primary political force behind the founding of KECC.

economic policies of heavy industrial development, planning and implementing not only large architectural and urban projects, but projects for the
chemical, steel, and shipbuilding industries. The KECC also functioned as
the organizational body through which the Japanese reparations and loans
were transferred to the development of the nation. Its largest projects were
the construction of Pohang Steel, at present the second largest steel-
manufacturing conglomerate in the world, and the creation of Kyungboo
Highway, linking the two largest cities, Seoul and Pusan, in Korea. Faced
with the immense task of developing the country without the appropriate
funds, the government had to perform the role of a broker whose primary
task was to bring in private capital to implement its plans. The KECC
was, in effect, its developer and technician; the implementer of its ideology
of progress and development. The Sae'oon Sang'ga and Yoido projects
thus stand at a critical juncture when the central government was attempting to pull in the concentrative forces of international and private capital.
They would showcase the future of wealth and prosperity promised by the
new powers. Not surprisingly, it is during this time that we see the emergence of new configurations in the discourse of culture, planning, and
modernism.

The utopian characteristics of the Sae'oon Sang'ga and Yoido plans
should be understood in this context. Though just four blocks were actually
built, the planners of KECC had intended the Sae'oon Sang'ga complex to
be part of an extensive redevelopment plan of the city centre of Seoul. Like
the Yoido plan, the plan projected a residential, office, and traffic complex
interlinked by pedestrian decks. The underlying concept of both projects
was that the city should be developed not along repartitioned chequer-
board lots but as an integrated whole through the input of large private
capital. The architects explained the logic of their project by first posing the
question 'what is the essential characteristic of modern civilization?'
The answer was 'the drive for development' which would be achieved
by the concentration of modern capital. 'For the CIAM principles of sun-
light, green and open space to be properly applied to Korea's real situation
we are squarely faced with the necessity that large amounts of modern
capital must be speedily invested and concentrated, from the stand point of
the public citizen, into the capital of our old Kingdom' (Korean Engineering Consulting Corporation 1969: 16). Thus the importance of the
image of the elevated deck crossing the boundaries of private property: a
realization of the integration of technology and enlightened capital. The
apparent acquisition of technical dominance, however, did not provide
the means of realizing the forms of their proposed city. The idea of integrated development overcoming the existing chequer-board grid results, as the
plans had stated, in the call for large-scale development funded by private
capital. Once these images proved to be utopian and unrealizable, all that
remained was the call for a development unhindered by culture.

What of culture then? As I have already mentioned, in the official discourse of the late 1960s and 1970s, it is made quite clear that while culture must be national, ethnic, and traditional, development must have a modern face; a policy akin to what Jeffrey Herf, in his study of Nazi Germany, has called 'reactionary modernism'. Witness the following statement by In Kook Chung, a juror of the aforementioned National Museum competition, explaining the rationale for the traditional forms of the winning design. 'We had ancestors full of wisdom and because it is a fact that the modern architecture of Korea cannot build and has not built anything as sublime as they, it is the imperative truth that we should devote our magical powers (engineering) in continuing the wisdom of our ancestors.'[10] The seeming grasp of technical command was then appropriated more specifically and more pragmatically into the regressive cultural policies of the Park regime. Science and engineering are considered to be basically formless and culture-free tools. Unlike Kim Swoo Geun's modernist urbanism which must express technology, for the traditionalist, it is a pure means in the continuation of a cultural heritage.

The discourse of culture and the discourse of development are thus separated. The consequence of this process is twofold. One is the uselessness of cultural forms in development, and the other a visual culture that is impoverished into a reified symbolic gesture. Absent is a dialectic that forces capitalistic development to face cultural conventions, thus enriching their possibilities. On the one hand, there is a discourse in which culture is a thing that can somehow exist on its own; reproduced, for example, in the form of a museum secluded in a palace, or in the forced preservation of residential neighbourhoods. On the other, there is a tortured modernism that allows itself to become the ideological face of development. Development is unburdened by cultural conventions and restrictions, leaving the utopian plans of the KECC to function ultimately and merely as pure ideology. This logic runs through proposals to erect alien monuments on sacred territories, to destroy ancestral temples to build national assemblies, and to tear down monuments to show that we have overcome the forces that built them.

The history of the transformation of Seoul has thus been a document of violent swings between destruction and ossified preservation, be it of existing relics or the reconstruction of traditional forms. This oscillation has many faces. It can be seen at the level of the planning ideology of individuals and institutions such as Kim Swoo Geun and KECC, and in the architectural forms of what were deemed cultural and modern. The most recent swing of the pendulum has unveiled the plan to dismantle the former headquarters of the Japanese colonial government in Kwanghwamoon. In August 1993, President Kim Young Sam announced that, in order to

---

[10] Quoted by Kim (1970: 9).

'restore the pride and spirit of the nation', the building should be speedily demolished. After the end of more than two decades of military rule the country continues to be caught in the logic of colonial symbolism.

More recently our understanding of the nature of modernity in Europe and America has grown from viewing it as a monolithic process of rationalization, bureaucratization, and the pursuit (now deemed futile) of a liberated society, to the understanding that it is constituted by contradictions, complexities, and conflicts. Scholars ranging from Marshall Berman to Michel Foucault have shown, in quite different ways, the immanent nature of the dualities that structured the modern world. In his *Condition of Postmodernity*, David Harvey characterizes the duality of the Western experience of modernity as the coexistence of 'destructive creation' and 'creative destruction' (Harvey 1989). Though the experience was obviously painful, there are many moments when a rich modernist culture had evolved through this dialectic. Kim and the architects of the KECC were modernists in terms of the architectural forms that they employed and also in the sense that the city was viewed as a problem as well as a tool for progress and economic development. Yet the KECC, with all its modernist language, was unable to link its plans to a specific ideological and cultural formation. Lacking was a concept of a modernist urban culture, i.e. an internalized, quasi-autonomous notion of what a modern city can be: aware of its specific economic and political conditions yet not totally subsumed within them. It has been the task of this chapter to show how these words have become a misnomer in the transformation of Seoul.

In the rapidly changing relation between concentration and centrality, the original projects by Kim Swoo Geun were ultimately unable to deal with the new economic realities of urban land. The Sae'oon Sang'ga complex stands alone in the dense urban fabric of the city centre, neither part of the surrounding area nor able to affect it. It is a fragment of a utopia that was reiterated in its full scale in the original plans for Yoido. The complex has thus proven incapable of functioning as a north–south axis of Seoul; nor has it facilitated the development of its adjacent areas as had been originally projected. The Seoul metropolitan government has in fact recently announced plans that the complex will be torn down in the year 2002, and in its place a pedestrian parkway is to be created. The everyday life that these environments sustain is obviously not as simple as the cultural and economic logic that produced them. One need not go into the tedious debate on the relation between the environment and its inhabitants to know that the intentions and forms of a plan by no means determine how people live in its realization. At the same time, the new urban forms test the idea of the multiplicity of landscapes deemed so characteristic of the built environment of pre-industrial Seoul. With the collapse of the social and physical structure of a medieval Hanyang, the nature of the landscape of a modern Seoul is a question that requires continued study. To understand the relation between culture and the built form of modern Seoul is to delve into the contradictions and dilemmas of modernism as it was formulated in the tragic history of colonialism, the Cold War, and military dictatorship.

References

Cho, Sung Yun (1992). 'Caste Structure and its Change in Seoul during the Late Yi Dynasty: A Historical Approach to the Emergence of Citizens'. Ph.D. dissertation, Yonsei University.

Harvey, David (1989). *The Condition of Postmodernity*. London: Basil Blackwell.

Jung, Inha (1996). *The Architecture of Kim Swoo Geun*. Seoul: Migunsa.

Jung, Kyung-Mo (1991). 'Park Jung-Hee: From Rise to Power to Tragic Fall', *History and Criticism*, Summer (13): 422–32.

Kim, Chae Yoon (1984). 'The Structure and Transformation of Korean Social Class', in *Theory of Korean Society*. Seoul: Minemsa.

Kim, Han Bae (1993). 'A Study on the Transitional Characteristics of Korean Townscapes'. Ph.D. dissertation, Seoul City University.

Kim, Kwang-Woo (1991). 'Daehan Jaekook and Urban Planning: The Urban Betterment Plans of Hansung', *Hyangto Seoul*, 50: 97–110.

Kim, Young-Mo (1971). 'A Study of the Formation and Transformation of Social Class during the Japanese Colonial Period', in Cho Ki-Joon *et al.*, *History of Everyday Life during the Japanese Colonial Period*. Seoul: Institute for Asian Studies, Korea University.

Kim, Won (1970). 'The Mirror of the Times: The Case of the National Museum', *Modern Architecture*, 3 (Sept.–Oct.): 8–10.

Ko, Dong Hwan (1993). 'The Development of Commerce in the Kyung'gang Area in 18th and 19th Century Seoul'. Ph.D. dissertation, Seoul National University.

Korean Administrative Research Institute (1980). 'The 3rd Chong'no Ave.–Daehan Theatre Redevelopment Project', *Case Studies in Korean Administration*: 200–13.

Korean Engineering Consulting Corporation (1969). 'Assumptions and Hypotheses for the Yoido Master Plan', *Space*, 29 (Apr.): 10–39.

Kyungsungboo (1928). *City Planning and Survey Report*. Seoul.

Lee, Jae-Sun (1979). *A History of the Modern Korean Novel*. Seoul: Hongsungsa.

Lee, Sang Koo (1986). 'The Process and Reality of the Formation of Urban Historical Space in Korea', *Urban Problems*, 21 (6).

——(1994). 'The Characteristics of Urban Form in Seoul during the Chosun Dynasty', Institute of Seoul Studies Seminar, 16 June (unpublished).

Lee, Tae Jin (1994). 'The Transfer of the Capital to Hanyang and the Demise of Fengshui', *Lectures in Korean History*, 14: 44–69.

Pai, Hyungmin (1985). 'A Study of the Genesis and Transformation of Urban Form: The Case of Yoido, Seoul'. Master's thesis, Seoul National University.

Rhee, Byeong Yul (1990). 'A Study on the Activities and Thoughts of City Planning in Seoul'. Ph.D. dissertation, Hanyang University.

Seoul Metropolitan Government (1977–83). *The 600 Year History of Seoul*, i–v.
——(1984). *Seoul Land Repartition Record*.
Seoul Metropolitan Government Han River Construction Center (1969). *Yoido and the Development Plan of the Han River Area*.
——(1971). *Yoido Comprehensive Master Plan*.
Sohn, Jung Mok (1990). *Studies on City Planning during the Japanese Occupation Period*. Seoul: Ilji Press.
Tange, Kenzo (1993). 'A Plan for Tokyo, 1960: Toward a Structural Reorganization', 1st pub. *Shinkenchiku*, Mar. 1961; repr. *Architecture Culture 1943–1968*. New York: Rizzoli.

7

Beijing: The Expression of National Political Ideology

Zixuan Zhu and Reginald Yin-Wang Kwok

More than 3,000 years old, Beijing has been the national capital of China for over 800 years. During the late period of Chinese feudalism, Beijing was the cultural and political centre of five dynasties—Liao, Jin, Yuan, Ming, and Qing. The unique culture and architecture of ancient Beijing had many striking characteristics. The city was constantly transformed under each dynasty out of elements from previous regimes. It also attracted numerous ethnic groups and absorbed elements from various nationalities, while at the same time sustaining characteristics of the dominant Han Chinese culture (Hou and Jin 1982). Under strong support from the national state, present-day Beijing has maintained many major traditional characteristics.

The People's Republic of China was established on 1 October 1949, and Beijing was designated as its capital city. This began a new chapter in Beijing's history. Rapid economic and political development since 1949 has not only expanded the city, but also made way for a new cultural and architectural environment. One example has been the evolution of Tiananmen Square from the front court of the feudal Imperial Palace to the focal point of national politics. Eighty-five years ago when China's last feudal regime, the Qing, was overthrown, Tiananmen Square was the entrance to the imperial Forbidden City, whose gates were closed to the public. Today, Tiananmen Square is a public space. Citizens and visitors arrive to witness the national flag-raising ceremony in the morning. People come to the square for their morning exercise, for sightseeing, and not infrequently to fly kites. Whenever there is a festival or holiday, Tiananmen is richly decorated and filled with crowds of spectators.

In Beijing as elsewhere, urban space has two types of function. As the physical setting for all human activities, it serves a utilitarian function. Urban forms are intended to satisfy the social and economic needs of the civil society. The built environment also serves a symbolic function, guiding public behaviour and expressing social messages. The national state plans the urban setting to inspire a political ideology, to enforce its legitimacy, and to demonstrate social morality. A city's physical environment is designed to fulfil both functions. Depending on the mix of the state–society relationship, the urban form usually reflects the strength and proportion of this dualistic dynamic. The relative emphasis given to the two functions varies in specific urban settings (Vale 1992; Kwok 1996). For example, a hospital, a factory, or a shopping mall serves a fundamental service need and thus has primarily a utilitarian function, although the physical design may carry some symbolic meaning. By contrast, a parade ground, a state capital, or a cultural centre, while serving a social need, is essentially an instrument of the national state, and thus has primarily a symbolic function, sending out implied political and regulatory messages in the expectation of a specific behavioural response. The first set of structures essentially provides a setting in which society functions, while the purpose of the second set is for the state to inform and to guide human behaviour.

Beijing's urban environment provides a historical record of changing political ideology and a stage for national political events. This chapter will review the city's political history and the impact of this history on the built environment. The historical development of Beijing is broadly separated into three periods: feudal, socialist, and future. The urban form of the various periods is viewed from the functional perspective in how the capital was utilized by the national state or served the civil society. As the national capital of both dynastic and contemporary China, Beijing was and is the locus of China's political power and state bureaucracy. It also projects China's national identity to the world. In its symbolic role, the city has been designed to reflect official political ideals, to confirm the legitimacy of the state, to reinforce the centralization of power, and to define the institutional relationship between the national state and the civil society. With a population of over 10 million, Beijing has to service this civil society and to provide a reasonably secure livelihood for its citizens. As the nation's capital, its utilitarian role of social provision through built environment becomes the model for the rest of China. Beijing's developmental dynamics periodically shifts in order to emphasize, adjust, or balance these two functions. In order to illustrate the historical meaning of Beijing's built environment, the centre point of the city, Tiananmen Square, is further examined in greater detail.

There are five parts in this chapter. Part 1 discusses Beijing as a dynastic capital and stronghold of feudalism, designed to serve China's imperial family and court. It reviews the traditional model of city planning and its successive applications. Part 2 discusses socialist Beijing as the seat of Chinese socialism and the manifestation of government policy. It examines the relationship between national developments as determined by the party and the state, and the principle of urban planning as applied to the capital city. Part 3 discusses contemporary urban planning for Beijing towards the twenty-first century, guided by the growth of the nation, the city's role in achieving China's national goals, and efforts to assure its future status as a global metropolis. Part 4 examines Tiananmen Square as an example of how layers of history and changing government ideology make their mark on urban form. The fifth part is a summary of the functional interpretations of Beijing's spatial environment.

Capital of the Feudal Era

Urban space and the built environment reflect a city's history and tradition, with each structure representing the period in which it was built and each period of the past contributing a layer of urban form (Kostof 1991). As development progresses, the built environment of earlier periods is partially or completely replaced by new structures that meet new needs. The neglect or demolition of past urban forms usually indicates rapid urban growth, with the emergence of new political and economic élites (Huxtable 1986). In contrast, preservation or renovation of old buildings normally signifies a more mature stage of urban development, with a well-established élite concerned with its historical heritage.

Chinese cultural traditions were based on Confucian thought and philosophy, and Confucian principles had a strong influence on how China's capital was planned and constructed. Confucians believed that ritual was of primary importance in ruling a country. Beijing was laid out according to the rules for a 'ruler city' (*wang cheng*) prescribed in the *Zhouli Kaogongji*, a planning and construction text based on Confucian principles (Fig. 7.1). One of the key principles is described below:

The artisan (*jiangren*) who plans the state capital, measures a square of nine *li* [Chinese miles] on each side with three gates on the sides. Within the capital, it is divided by a grid of nine paths running both horizontally and laterally. The north–south streets are nine carriage tracks (*gui*) in width. On the left [east] is the Imperial Ancestral Temple, and to the right [west] are the Altars of Soil and Grain. In the front [south] is the outer court and behind [north] the marketplaces. (He 1985: 29–31)

Beijing was laid out according to a strictly prescribed hierarchical arrangement. The Imperial Palace was positioned at the centre, with the Imperial Ancestral Temple to the east, the Altars of Soil and Grain to the west, the 'outer court' (*wai-chao*) and government offices to the south, and market places and space for commoners to the north. Both Beijing's original plan and its subsequent development during the feudal period were based on this ideal model. The ideal of *lizhi* (use of ritual action to rule a country) reinforced the feudal class system and established the sanctity of imperial rule.

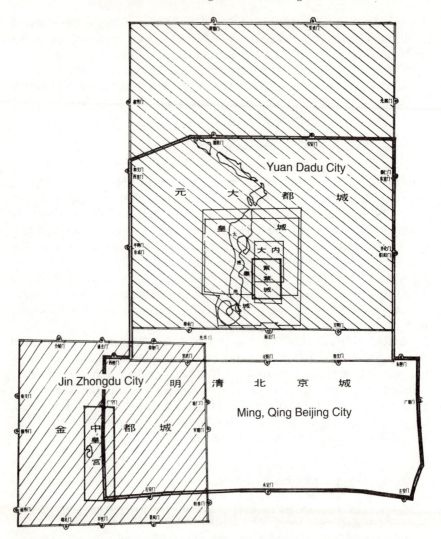

FIG. 7.1. Model plan for a capital city during the feudal era in China

According to historical record, during the Jin dynasty, the city of Zhongdu was located slightly to the south-west of present-day Beijing. In 1215, the Jin dynasty was overrun by Mongols, and Zhongdu's central palace was destroyed. A second palace (Daning) in a north-eastern suburb was saved by the invaders. With a large area and an ample water supply, this undisturbed part of Zhongdu was selected by the first emperor of the Yuan dynasty, Kublai Khan, as the location for the new city of Dadu in 1267 (Fig. 7.2).

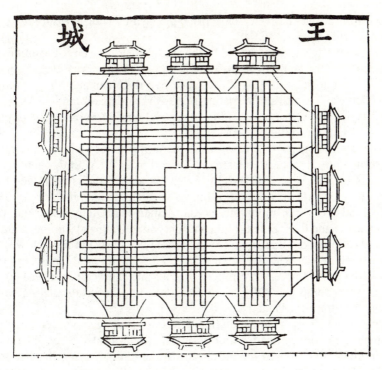

F IG . 7.2. Changes in feudal Beijing

The Confucian ideal of *lizhi* imposed a clear and strict feudal class system. Dadu was planned with the Daning Palace (today's North and Middle
Lakes) at the centre of the urban landscape and water system. The palace
and the park provided a unique physical environment in the capital's central district.

During the early Ming dynasty, the emperor moved the capital to Nanjing
and gave Dadu to his fourth son. After staging a *coup d'état* in Nanjing, the
fourth son moved the capital back north to Dadu because he considered the
city his 'dragon flourishing (*long xing*)' spot. He reinforced the city to
combat a southward invasion by the Mongolians, and renamed his capital
Bejing. The imperial walled city was moved further south, away from the
desolate northern part of the former Dadu, also to accommodate the expansion of the front court (*qianchao*), reflecting the palace form of Nanjing
(Fig. 7.2). The rebuilding of the Ming capital included the construction of
the 'Thousand-Pace Corridor' (the present Tiananmen Square) in front
of the Imperial Palace with government offices lining the east side and
the military headquarters lining the west. An artificial hill, *Jing Shan*, was
built north of the palace. The city as a whole was laid out in a rigid, orderly

fashion based on the 'ruler city' principle, symbolizing the supremacy of imperial rule.

An outer city was later added to the south side of the city. Together with the existing city, they formed a double-square shape which was unique at the time (Hou and Jin 1982). Taken together, the Yuan and Ming dynasties provided Beijing with a firm spatial framework and major landmarks, giving the city its strong formal identity and regulating its subsequent physical development up to the present.

The Qing dynasty that came next made no major changes but rather preserved the city inherited from the Ming (Fig. 7.3): 'After designating the new capital, the palace and the city layout remain[ed] the same' (Chen 1977: 38). The greatest contribution of the Qing rulers was in landscaping. They renovated the 'three mountains' (Xiang Shan, Yu Quan Shan, and Wan Shou Shan) and five gardens (Jing Yi, Jing Ming, Qing Yi [Summer Palace], Chang Chun, and Yuan Ming [Royal Summer Palace]) in a north-western suburb. These were, however, all reserved for the imperial family, but they sparked a trend among government officials and the wealthy. These groups began to develop private villas and gardens that have left Beijing with a valuable cultural heritage.

Under the Yuan, Ming, and Qing dynasties, Beijing served as a national political and cultural centre and an international centre of exchange and communication. Toward the latter part of the Qing dynasty, the city accumulated cultural inputs from different Chinese ethnic groups and from the West. The most prominent imports from the West were in the areas of science and technology, religion, customs, and entertainment. The private sector contributed a myriad of public built forms. The construction of Peace and Harmony Lamasery (Yonghe Gong), the mosque at Ox Street (Niujie), the Confucian Temple, and the National Academy demonstrated the city's ethnic diversity. The Western presence was exemplified by the embassy district (Dong Jiaomin Xiang) (Arlington and Lewisohn 1987) and the Catholic church. A new broad-based capital culture (*jing wenhua*) emerged that included an assortment of dialects, folk customs, traditional operas, national costumes, ethnic recreational activities, and religious rituals (Hou and Jin 1982).

Following Confucian tradition, Beijing was planned along a central north–south axis, representing the authority of the state. The Imperial Palace, government offices, religious buildings, and minor royal residences were all located, often symmetrically, on the east and west sides of the central axis. Political power and social position were clearly demarcated in the urban landscape. Western colonial powers began invading China in the 1830s, but their economic interests were concentrated mainly on the coastal seaports. Beijing was under considerable political pressure from the West, but few colonial buildings were erected, and these were limited to embassies and educational institutions. The urban form of the feudal capital was left intact.

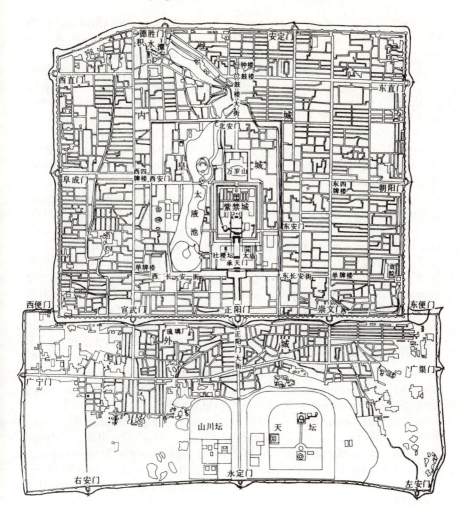

FIG. 7.3. Beijing during the Ming and Qing dynasties

Capital of the Socialist Era

The contemporary urban physical environment is largely a product of economic development. As the industrial and post-industrial economy grows, modern structures increase in number and improve in quality. Built forms also become more complex and varied in response to economic and social diversification, reflecting a growing number of groups with varying needs and demands. Cities tend to pass through a period of unstable and mixed development as construction catches up with an expanding economy and rising social expectations. Competition between social amenities and state enterprise leads to spatial conflicts and a shortage of urban land. As this competition becomes more severe, land use becomes intensive: high-rise structures become the norm (Harvey 1989: 211–83), and land competition soars. In this process, the civil sector usually lags behind the production, particularly in socialist nations. Under these circumstances, urban growth is accompanied by congestion and inadequate public amenities—the more rapid the growth, the more complex and serious the land shortage.

Soon after the proclamation in October 1949 of Beijing as China's capital, the new government began the task of transforming this ancient city into a modern socialist metropolis (Hou and Jin 1982). During the early stages of nation-building, China's primary mission was reconstruction and industrialization. The Rehabilitation and the First Five-Year Plan (1949–57) periods concentrated on economic development based on large-scale heavy industry. The primary goals were national independence and economic growth. Centralization and consolidation of state power were considered necessary for nation-building on both ideological and developmental grounds. The slogan 'the nation as a chessboard' characterized the mobilization of resources and labour to usher in a socialist transformation (Riskin 1991). At the national level, the commitment to industrialization meant an emphasis on developing physical infrastructure, human capital, and administrative systems. Industrialization also generated intense pressure to urbanize. Urban planning and construction focused on production and administrative requirements, and feudal and capitalist 'consumer' cities were criticized as being exploitative (Kirkby 1985).

For Beijing, priorities included developing infrastructure, improving health conditions, and relieving the housing shortage. The Beijing Urban Planning Committee, established in 1949, with Chinese and foreign experts, made many planning proposals for the city's development and expansion. Beginning in 1953, China initiated a construction programme that included several large-scale projects in Beijing.

As early as the 1950s, the leading Chinese architect, Liang Sicheng, suggested that a new administrative capital be built in the western suburb of Beijing in order to preserve the entire historical cityscape (Liang 1986). However, the old city occupied 62 square kilometres and held a population of nearly 1 million. To maintain the traditional capital as a museum was considered to be too costly and unmanageable. Instead, a plan which expanded modern development outward concentrically from the historical centre was adopted, but it created an acute spatial conflict between the old and the new.

Both the central and the local governments considered urban planning important. Proposals made in 1953 and 1956 reflected the major principles guiding city planning at the time: serving production, serving the state, and serving the workers. The main goal was to rebuild Beijing as the political, economic, and cultural capital of the nation, with emphasis on creating a science and technology centre with a strong industrial base (Chen 1977). A new city plan was developed that designated the city as the centre and expanded development outward in a radiating pattern with corresponding transportation and communication systems. The plan was targeted to accommodate a population of 5 million within an urban area of 600 square kilometres.

Beginning in 1954, Beijing initiated the practice of land-use zoning. Several industrial districts were developed in the north-eastern, eastern, southern, and western suburbs. The north-western district was reserved for institutes of higher education and scientific research. National institutes and government offices were retained in the city centre and in the western suburbs: these included the city's administrative headquarters. The residential districts were developed gradually, street by street and block by block. In general, the physical environment of modern Beijing was a combination of the utilitarian mode of social and economic planning and the response to the symbolic need of socialist nation-building. The capital was to be distinctly different from both the feudal ancient capital and the industrialized city of Western capitalism.

The Great Leap Forward began in 1958. This was largely a period of decentralized rural development. The slogan 'walking on two legs' referred to the simultaneous adoption of methods that were both traditional and modern, indigenous and foreign. It applied to production technology and to other spheres such as art and culture. The guiding rule of development was achieving a balance between contradictory forces. Development and administration were decentralized to the countryside, but policy decisions remained firmly in the hands of the central government. The rural commune was introduced, not only to stimulate agricultural production but also to serve many non-farm functions in the rural areas. These included social services, military activities, and industrial production (Riskin 1991). The process of rural development was designed to create many small, low-density towns (Kwok 1981).

Urban planning in Beijing was significantly revised in line with the new ideology. One key principle was the elimination or reduction of the 'three differences'—between workers and peasants, between urban and rural, and between intellectual and physical labour. Committed to dispersed development, planners proposed a 'scattered collective' model for cities to parallel the rural communes. Applied to Beijing, the new urban model had four objectives. The first was to spread the city into an urban region, which would be twenty-eight times bigger than the planned inner city, but to maintain the same population target. Second, the population of the inner city was to contract from the existing 5.0 to about 3.5 million—equivalent to a 30 per cent reduction in population. Third, the city was to be decentralized into a number of 'scattered collectives', separated by agricultural land or

green belts. Fourth, the plan provided a generous amount of parks and
open space: 40 per cent of the inner city and 60 per cent of the suburbs were
to be converted into park land (Beijing Jianshe Shishu Bianji Weiyuanhui
1987). Although this ambitious plan was never fully implemented, Beijing
region nevertheless became more dispersed and rural than concentrated
and urban in character (Fig. 7.4).

These planning principles were similar to the 'organic decentralization'
theory advocated by Eliel Saarinen (1943) in Europe. They certainly had
the potential to achieve environmental benefits, but, perhaps more importantly, they represented a radical attempt to unify the life-style, labour
practices, and social behaviour of China's large rural and urban populations. Urban planning would provide the spatial conditions necessary to
achieve these idealistic goals. Within a few years, these goals were discard-
ed, but the engendered spatial concept still provided the basic principle for
planning in Beijing.

During the Great Leap Forward years (1958 to 1961), industrial development continued in Beijing. More significantly, several major construction
projects were carried out to celebrate the tenth anniversary of the new
nation. In the city centre, Changan Street was widened and extended, and

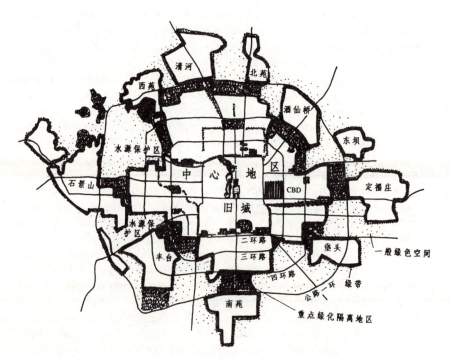

FIG. 7.4. Beijing during the Great Leap Forward: 'scattered collectives'

Tiananmen Square was expanded and rebuilt. The 'Ten Great Architecture' projects, including the Workers' Gymnasium, the Beijing railway station, the People's Great Hall, and the Museum of History and Revolution, were all public building complexes (Wang 1959). Together, they were to enhance the symbolic significance of the capital, celebrate the nation's achievements, and mark the starting-point for a new era of spatial development in Beijing.

Planners devoted considerable attention to architectural style and urban ambience. Many studies focused on how new development projects could complement Beijing's traditional character and fit into the city's historical framework, giving rise to many controversies. Some buildings constructed in the early 1950s had simply copied prominent features of traditional architecture: the large pitched roofs were particularly popular. By 1954, this approach was criticized by a political movement which opposed old traditions and unnecessary waste, as well as the indiscriminate copying of modern Western architecture.

These debates were primarily ideological. Traditionalism was equated with a revival of feudalism, whereas modernism was equated with an adoption of capitalism, and both ran counter to political goals. In 1958, the government proclaimed an architectural style for the new China called 'Chinese also new' (*zhong er xin*), which included historical Chinese characteristics but reflected the new socialist spirit (Samuels and Samuels 1989). A design contest in Beijing confirmed the 'Chinese also new' approach, and many of the public buildings constructed for the tenth anniversary of the People's Republic followed the new style. The 'Ten Great Architecture' projects are prime examples. These monumental buildings were stately and imposing; however, they were also more in keeping with traditional imperial architecture than most buildings constructed during the previous ten years. In this short period, the project selection as well as the building design demonstrated the height of architectural symbolism in socialist China.

The Great Leap Forward was followed by the period of Economic Recovery. Although agriculture was reinstated as the foundation of the economy, industry was still the leading sector of development. Political and administrative control was once again centralized, and ideology was explicitly included as a key component of the planning process. Once the economy regained momentum, the nation shifted its attention from production to politics. During the Cultural Revolution that followed (1966 to 1976), ideology took precedence over production, resulting in considerable economic disruption. 'Serving the people' was now the motto, with production planned primarily to meet social needs. Industry was decentralized but remained under centralized leadership (Riskin 1991).

Urban development generally slowed down during the Cultural Revolution, but Beijing was an exception. Steady population growth brought

increasing pressure on housing, transportation, and environmental resources, while heavy industry increased the demand for water and energy
and contributed to environmental pollution. In short, Beijing experienced
all the typical problems of a fast-growing city.

During this period, however, the government embarked on a programme
to improve Beijing's public facilities. A programme of public facilities for
civil society included the construction of many new district libraries and
museums, plus hospitals, theatres, and cinemas throughout the city (Beijing
Jianshe Shishu Bianji Weiyuanhui 1986). Another programme was
launched to build a public gymnasium in each urban district and suburban
county. These additions enriched Beijing's social and cultural life and
helped transform the city into a more hospitable one for its citizens. The
planning policy was, primarily, to facilitate the capital's civil society.

The Cultural Revolution was followed by a shift toward a market economy that included an open policy to the outside world (Riskin 1991).
China entered a period of economic reform with an open-door policy. This
new approach encouraged urbanization and led to a greater level of inter-
nationalism. It also entailed a number of adjustment problems. The impact
was initially confined to the production sector but quickly spread to the
cultural and social arena.

By the late 1970s, around the embassy area in east Beijing, a construction
explosion boasted new high-rise apartments, hotels, offices, commercial
plazas, and recreational facilities for foreign visitors and the international
community. In April 1980, the Central Secretariat of the Communist Party
proposed a new set of policy guidelines for construction in the capital city
(Beijing Jianshe Shishu Bianji Weiyuanhui 1987). They specified that Beijing's primary function would be as a national political and cultural centre
and China's window to the outside world. The city's industrial function was
now deliberately eliminated. Rather, the new guidelines emphasized the
improvement of Beijing's social and environmental conditions, the development of the city's educational institutions and scientific and cultural
facilities, and the promotion of economic prosperity based on the service
sector.

In response to the new directives, Beijing's overall plan was revised
between 1980 and 1982. The new plan emphasized the city's role as a
famous historical and cultural centre and specified that urban planning must
reflect Beijing's Chinese history, indigenous cultures, revolutionary traditions, and unique social characteristics (Beijing Jianshe Shishu Bianji Weiyuanhui 1987; Beijingshi Tongji Ju 1984). The State Council approved the
revised plan in 1983. The architectural dictum of 'Chinese also new' was
once again in full force. Economic reform brought in a major shift of
emphasis. The built environment was to glorify and consolidate the importance of the national state not only for its own citizens but also for the
international community.

Since the 1980s, Beijing has experienced an unprecedented construction boom. Sites for new hotels, offices, and residences are in high demand, and old buildings are torn down for redevelopment. Construction is funded largely by the private sector, fuelled by foreign capital. As the population has increased and land-use strategies have shifted, new demands have heightened land competition. The most urgent problems have emerged in the central districts where modern developments are springing up. Here, new buildings often contrast sharply with the old, and the results are not always visually harmonious. Such aesthetic conflicts, however, are not the only problem. More serious is the increasingly intense competition for limited space. Conflicting demands for land add a new and unfamiliar component to Beijing's urban planning strategies.

In an effort to keep up with this level of growth, the city has invested in infrastructure development. Major water-treatment facilities and a large national library were added in the suburbs. Highways now extend concentrically from the Second Ring Road to the Fourth Ring Road, and an interurban super-highway links Beijing, Tianjin, and Shijiazhuang. The National Olympic Sports Centre and Asian Olympics Village were constructed for the 1990 Asian Olympics (Cui 1990a, 1990b), an event that helped promote Beijing's image as an international city. The site for this building complex was particularly prominent—on the central axis and north of the Imperial Palace. Urban lanscaping continues to be a priority, but the focus has shifted away from large city parks to small neighbourhood parks and open spaces, giving the city a more amiable appearance. Apart from their utilitarian functions, this new construction, landscaping, and infrastructure serve to bolster Beijing's image as a national hub and an international centre.

With these many functional changes from symbolic to utilitarian and vice versa, and in spite of the many efforts to improve the society, the public facilities built for citizens were scattered and inserted into the residential neighbourhoods. Their appearance was neither distinctive nor enticing. These civil constructions were merged into the urban fabric and blended well into the cityscape, thus losing their spatial impact. Beijing, as the locus of the national state, exuded its political significance. Its urban form was still dominated by the Imperial Palace and the new nation's political complex at the spatial pivot of the capital, located at the intersection of the city's north–south and east–west axes. Here was Tiananmen Square, the symbolic site of the central state through many successive governments.

Capital for the Future

The global economy has a particularly strong effect on East Asian development. Linkage with transnational capital networks is one of the major sources of economic growth and industrial restructuring (Kim and Kwok 1991). Because of its effectiveness, global production is accepted as a model for development. Many Third World nations have utilized foreign direct investment as a effective tool for economic development. This mode of national development, however, has produced specific urban and regional growth and spatial patterns, altering the local spatial and cultural environment into an international indigenous mix (McGee 1991). Since the period of economic reform, China has adopted the same strategy to induce growth. The composite cultural and spatial effects are most pronounced in south China where foreign investment first entered (Kwok and Ames 1995). Accordingly, Beijing, in planning for the twenty-first century, prepares its spatial environment, ready for the drive for transnational capital.

The growth of a socialist market economy in the early 1990s brought with it a number of major institutional, political, economic, social, and cultural changes. As the capital, Beijing was the center of change, and spatial developments in the city have reflected many policy experiments. In response to the rapid rate of urban construction, the opening to the international economy and the changing political climate, the city government formulated anew the Beijing Overall Urban Plan in 1991 to meet anticipated needs up to 2010 (*Beijing City Planning and Construction Review* (1993); Shoudu Shehui Jingji Fazhan Yanjiu Suo 1989). The plan covered five major areas.

First, as the national economy is to be fully open to the outside world, the city is to develop the service sector and new and high-technology industries. Second, the designated function of Beijing, in addition to its previous function as the nation's political and cultural centre, is to be a modern city and an important historical centre, to be known internationally. Third, the urban population is to be restricted to 15 million by 2010, with a permanent population of 12.5 million and a transient population of 2.5 million. The peak permanent population as of 2040 is to be no more than 14 million. Fourth, Beijing's inner-city area is estimated to expand by three-quarters of the 1956 proposed size or about 1,040 square kilometres. Growth is to be distributed to the expanding satellite towns within the urban region. The city centre is to be extensively redeveloped in order to accommodate an international financial and business district in the eastern part of the city adjacent to the existing embassies and foreign community. The three existing shopping centres—Wangfujing, Xidan, and Qianmen—have been planned for substantial redevelopment and regulation. Other new commercial and cultural centres would be positioned in a multinuclei pattern to service the increasing residents. Fifth, preservation of historical and cultural heritage has been emphasized. The old city is to be preserved in its entirety, in particular, maintaining the central axis, the water system, and landscaping. This recent plan was supplemented by a comprehensive list of requirements on infrastructure, communication, housing, and landscaping, aiming at internationalizing the capital.

Beijing's construction boom has continued over the past few years. According to the new plan, key infrastructure development included the western railway station for domestic travel, and a new direct highway to the international airport facilitating international travel. Tourist services such as the International Convention Centre, the International Silk Boutique, the World Park, the Chinese Nationalities Park, and the Old Beijing Miniature Garden have been built around the city, mostly along the northern part of the central axis. Foreseeing a burgeoning of economic growth, the city concentrated its development focuses on the new international financial and business district and its central shopping centres. The proposed new international district is now under construction, and two of the central shopping centres, Wangfujing and Xidan, are being renovated. As foreign investment has been pouring into the capital, a surge in property development has been occurring, most obviously illustrated by the international hotel complexes appearing in the city centre surrounding the Imperial Palace. In addressing the city's needs for the twenty-first century, planners are developing new spatial strategies to combine the attractions of an ancient capital with the requirements of an international post-industrial metropolis (Fig. 7.5). To avoid visual clashes, the plan locates the modern structures away from traditional landmarks, separating the historical layers of urban form.

While allowing for economic diversification and internationalization, the national state under socialism continued to stress political unity, social stability, and an unswerving commitment to manage economic growth under monolithic, hierarchical state bureaucracy. In addition to constant exhortation, highly visible public buildings and monuments serve as symbols and reminders of these priorities. In historical terms, the introduction of socialism represented a radical departure for the nation, fundamentally alien to traditional feudalism. The new system had to be instilled throughout society, allowing only those nationalistic traditions to remain that in some way enhanced the current ethos. Elements of the built environment have served as important symbols to consolidate the new political culture and to instruct citizens on correct behaviour. In this context, older structures have been retained as reminders of national identity and continuity.

At the same time, urban space is required for the new structures that inevitably accompany economic growth. As international and domestic capitalists began to construct their imposing buildings, the government worked to keep the new edifices as much as possible in a subsidiary position. Yet the new built environment, motivated more by utilitarian requirements than by political symbolism, is likely to assume greater importance in the future. Thus the contest for urban space will be between state bureaucrats and private entrepreneurs: the outcome will be determined not by

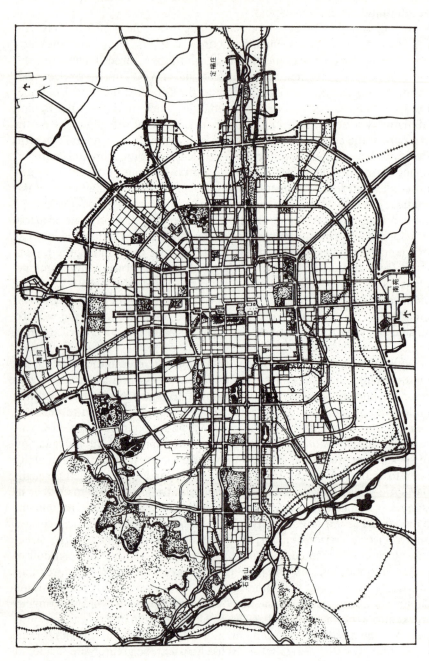

Fɪɢ. 7.5. Beijing in the early 1990s

zoning allocations in the city planning office, but by debate and decisions in the political arena.

The Evolution of Tiananmen Square

The spatial environment of the capital has been receiving special consideration in Chinese urban planning and construction since the feudal era. A national capital must be magnificent, with an air of dignity and authority (Liu 1989). The state's administrative complex should be in the centre of the capital. In Beijing, nothing illustrates the historical continuity of this principle more clearly than the long history of Tiananmen Square. Located in front of the Imperial Palace, the square represented the supreme sanctity of feudal imperial rule, and the state's monopolistic power over the society. In modern China, it plays similar roles. Throughout Tiananmen's history, its design has been consistently following the Confucian principles.

Tiananmen Square in the feudal era

Tiananmen Square followed strictly the traditional 'ruler city' design, though it evolved through several dynasties. Part of its origin was from Changan. The capital city of the Tang dynasty provided a model for urban design. It was the largest capital of the early feudal period, and its layout was the most orderly. The city's clear central axis represented imperial rule and provided a reference for spatial planning. In front of the south palace gate an enclosed courtyard (front square) extended in an east–west direction. This courtyard was the site of imperial feasts, celebrations, ceremonies, announcements of amnesty, and imperial receptions for foreign envoys (Hou 1979). Another part of the square's origin was from Bianliang. In this capital of the northern Song dynasty, the front square stretched in a north–south direction. On this axis was the Imperial Way (*yudao*), lined with imperial corridors on both sides (*yulang*). Together, they formed the 'Thousand-Pace Corridor'. This corridor was partially opened to the common people. During festivals, various activities were held in this courtyard for the public. The predecessors of Beijing, Zhongdu (capital of the Jin dynasty) and Dadu (capital of the Yuan dynasty), both copied the design of Bianliang's 'Thousand-Pace Corridor' arrangement of a north–south axis (Hou 1979).

In Jinling (present-day Nanjing), capital of the early Ming dynasty, the front square combined the east–west alignment of Changan and the north–south alignment of Bianliang to form a 'T' shape. When the Ming capital moved to Beijing, this design was adopted. During the Ming and Qing dynasties, the front square was moved further south, outside the Imperial Palace, to the site of the present Tiananmen Square. Administrative offices of the central government lined both sides of the 'T', forming what was collectively called the 'outer court' (Fig. 7.6) (Hou 1979; He 1985). Use of this area was restricted to officials and the imperial family: it was completely closed to the public, heavily guarded, and off limits for the common citizens. Outside directly to the south was a chequer-board of commercial and shopping streets, a popular public gathering place. This spatial arrangement clearly reflected the absolute separation of the state from society during this era.

During the feudal period, Tiananmen Square was the site of a number of state ceremonies. Here, civil and military officials gathered before dawn to present themselves to the emperor. Emperors and empresses were crowned here. Imperial edicts were proclaimed from the top of the palace wall, and the written document was lowered into the square where a state official received and pronounced it to the public. This ceremony illustrated the hierarchical order of the nation and the segregation of the emperor, the state bureaucracy, and civil society. Other major annual events held at Tiananmen Square included the announcement of the results of the highest civil service examination, and the start and finish of the horseback parade of the three top candidates. This signified the state's admittance of selected members of society into the official bureaucracy. Just outside the west gate, open trials were held in the autumn for the most serious crimes (Hou 1979);

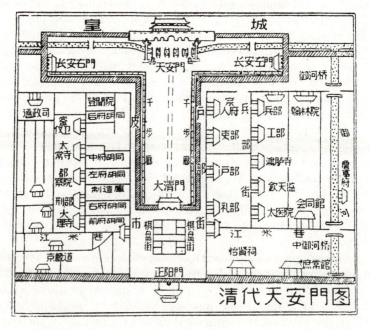

FIG. 7.6. Tiananmen Square in the Qing dynasty

thus, it was the place of supreme justice. All legislative, executive, and judiciary functions of the state were concentrated in this space.

Tiananmen Square in the twentieth century

The revolution of 1911, led by Dr Sun Yat Sen, put an end to 1,000 years of imperial rule. For the first time, citizens were allowed into Tiananmen Square. In 1913, the southern area of the Imperial Palace with three major pavilions, called the front court (*qianchao*), was opened to visitors. The Forbidden City was partially and gradually transferred into the Palace Museum. The 'May Fourth Movement' in 1919 started in Tiananmen, and established the square as the civil political as well as the cultural centre of Beijing.

During the republican period, the square went through several changes. The 'Thousand-Pace Corridor' was torn down, and gates were removed. Streets were connected through the square; consequently, parts of the wall had to be demolished to accommodate the through traffic and to reduce congestion around the square. These alterations diminished the square's original concept of tight enclosure and exclusion—it had become a social space (Siren 1985). In 1945, at the end of the Second World War, Tiananmen Square teemed with civil and political activities. Several large-scale student demonstrations took place here, such as the 1947 anti-hunger and anti-civil war protests.

The declaration establishing the People's Republic of China, on 1 October 1949, and the celebration of the birth of the new nation both took place at Tiananmen Square. The square had entered a new era: it was now the national political centre; moreover, its main gate was adopted as the national emblem, the symbol of a new China. Over the next ten years, Tiananmen Square went through many more changes. By connecting Changan Street through the northern edge of the square, this main street and the square were integrated. Instantly the square was the most important node of the east–west traffic in Beijing. All parades marched down this road. In 1955, by removing the traditional administrative offices, Tiananmen was expanded to the east and west. Even with this extension, however, the square was not large enough to meet the new demands of national celebrations and festivities. For example, during the annual national day gathering and parade, an estimate of more than 1 million people lined up along Changan Street.

As part of the preparations for the tenth anniversary celebration of the People's Republic of China in 1958, the state decided to expand Tiananmen further by enclosing it with several large public structures. The government organized a national design competition for the square and solicited proposals from the public. Summarizing more than thirty proposals, they were combined into a master plan (Beijing Jianshe Shishu Bianji Weiyuanhui 1987). The function of the square was designated as political. It was enlarged to form a rectangle measuring 500 metres east to west and 860 metres north to south, creating the world's largest square, to be used primarily for public ceremonies. Changan Street would be further widened to accommodate national parades, and to form an east–west linear extension of the square. The traditional central axis was to be maintained and enforced with the Tiananmen gate as its northern focus, flanked by two new complexes; the Great Hall of the People on the west and Museum of Chinese History and Revolution on the east. Both buildings were designed on the 'Chinese also new' principle. A national flagpole and a Monument to the People's Heroes would be located at the centre as focal points. The monument was the product of six years of consultations with people throughout the nation: it was inscribed with calligraphy of Mao Zedong and Zhou Enlai.

After the completion of this plan, the square became even more spacious and more imposing (Fig. 7.7). The north–south axis was intersected by a new east–west axis formed by Changan Street. Tiananmen Square was now spacious enough to accommodate national festivals and parades. The new spatial arrangement represented both the ideological contrast between the feudal past and the socialist present and the historical continuity of the nation. Overall, the square was a powerful expression of the political objectives (Zhu 1989).

In 1977, a mausoleum was built for Chairman Mao in the southern part of the square one year after his death to commemorate the ideological leader and the founder of the nation (Cohn and Zhang 1992; Kwok and Kwok 1979). This addition further strengthened the square's role as a political and historical centre. The significant memorial was an immediate popular destination for both domestic and international visitors (Fig. 7.8) (Chen and Sun 1989). During holidays, the area teemed with crowds of people.

Beginning in the mid-1970s, Tiananmen Square also revived its political role as a place for confrontation between the state and society. Two months after Premier Zhou Enlai's death in 1976, a spontaneous civil demonstration of mourning (BBC 1978) for this selfless and just leader marked the beginning of the Anti-Gang-of-Four Movement. This event also heralded a major leadership change and the launching of China's Economic Reform. In April 1989 the death of Hu Yaobang, a leader for democratic reform, sparked another civil demonstration (*Beijing Review* 1989). These in turn led to a drastic state intervention—the Four June Tiananmen incident (Hung 1991). Thus the square has provided a spatial setting for political acts that have had serious and far-reaching effects on the entire nation. These recent public encounters between the state and society have established Tiananmen Square as the place where contemporary political acts take place. Amid the historic buildings and contemporary monuments, new ideologies challenge the old, and the citizen meets with the bureaucracy. Perhaps most significantly, it is turning into the primary locale where political

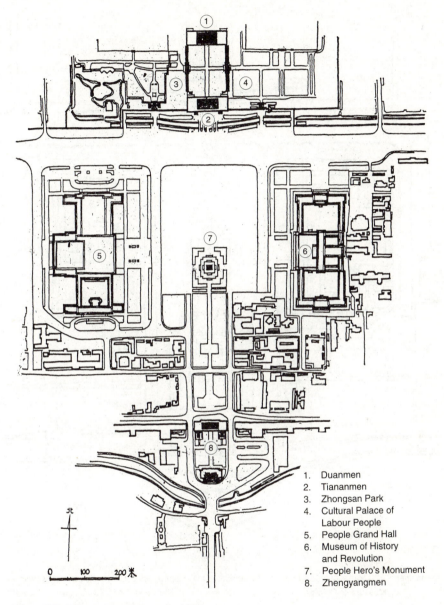

1. Duanmen
2. Tiananmen
3. Zhongsan Park
4. Cultural Palace of
   Labour People
5. People Grand Hall
6. Museum of History
   and Revolution
7. People Hero's Monument
8. Zhengyangmen

FIG. 7.7. Tiananmen Square in 1959

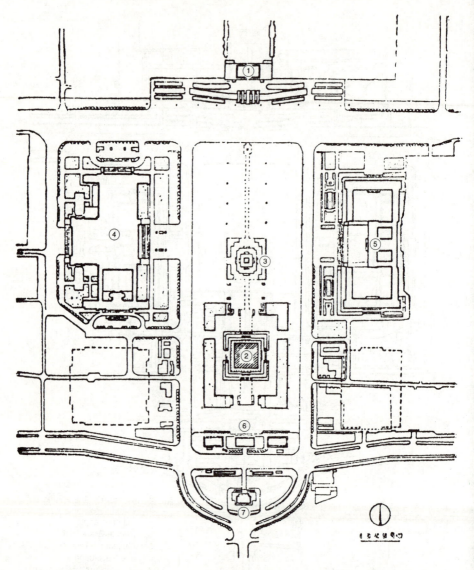

1. Tiananmen   2. Mao's Mausoleum   3. Monument to the People's Heroes   4. Great Hall of People
5. Museum of Chinese History and Revolution   6. Zhenyangmen   7. Jianlou

FIG. 7.8. Tiananmen Square after 1977

ideas and ideologies are openly voiced and exchanged (Hershkovitz 1993).

Positioned at the centre of Beijing's traditional north–south axis and the new east–west axis, Tiananmen Square is the heart of the capital. Surrounded by carefully selected and designed structures, the square has taken on immense political and ideological meaning, symbolizing not only the authority but also the historical continuity of the state, with the imperial dynasty replaced by the socialist republic. The urban transportation system radiates from the square, ensuring that national events and debates throughout the capital are quickly communicated to this central exchange point. As Beijing moves toward the twenty-first century, Tiananmen is truly at the nation's centre, in both symbolic and utilitarian terms.

Conclusion

In feudal Beijing, the spatial environment was designed almost entirely for the imperial family and the ruling élite. There was little concern for the population at large other than providing the necessary infrastructure for transportation and water supply. There was also little concern for the outside world: the urban physical environment reflected a state ideology that was essentially inward looking.

One cardinal principle of city planning was to facilitate state control of the citizens: no open spaces were provided that would encourage public gatherings or interaction among different social groups (Liu 1989). The physical environment was designed to display both power and permanency. The state built structures in strategic locations to form a powerful network of landmarks that defined the spatial characteristics of the capital and guided its growth. This was essentially a 'ruler city'—a capital that met the needs and expressed the ideology of the state.

In modern times, the government has devoted considerable effort to developing the economy and to improving the welfare of the population, but policy-making and implementation have tended to remain the monopoly of the state. Under early state socialism, public policy focused more on production than on welfare, although basic infrastructure and social services were improved. Another priority was the creation of a whole new government apparatus and set of public institutions. In Beijing, modern buildings were designed to house the essential state functions and, perhaps more importantly, to affirm and glorify the new regime. The permanence and grandeur of these structures represented the permanence and grandeur of the state, the leadership of the party, and the authority of the administration. As China has become more open to the international community, the city has also come to serve as a showcase for national achievements and a window for foreign investment.

The major urban design question in Beijing today is how to reconcile the architectural styles and land-use patterns of feudalism and socialism while at the same time expressing the continuity of Chinese history and culture. As in the 1950s, the design principle 'Chinese also new' is considered particularly appropriate. The objectives are to excel in the global community without losing China's national identity.

The historical development of Tiananmen Square illustrates a gradual trend toward secularization, at least in the utilization of space. After feudal times, the square was been expanded, and the barriers surrounding it have been removed. It started as an imperial sanctuary and command centre, developed into an open-air museum displaying the life-styles of past emperors, and eventually became a truly national space, used by both the state and society as well as international visitors. Throughout this evolution, an ever wider cross-section of the population has come to use the square for an ever wider range of activities. As the central place where major proclamations have been made and major events have taken place, Tiananmen Square provides a sense of physical continuity between the past and the present, symbolizing the essence of 'Chineseness'.

Today, the large number of people who come to Tiananmen Square and the variety of ways in which this central point is used may provide a source of conflict. As the symbolic centre of the nation, the square also has immense significance in the media. Ideological views are expressed here, and both the state and society air their agreements and disagreements. The open exchange and challenge mark a new era and function for the square. Thus, the historical evolution of Tiananmen Square illustrates the transformation of Beijing's spatial environment—from a symbol of state power to a domain of civil society. Throughout its long history, the square has expressed government ideology and now occasionally civil opposition. Today it represents an accumulation and a chronicle of Chinese political history.

References

Arlington, L. C., and Lewisohn, William (1987). *In Search of Old Peking.* Hong Kong: Oxford University Press.

BBC (British Broadcasting Corporation) (1978). *Summary of World Broadcasts, Part 3.* FE/5977, 24 Nov.

Beijing Jianshe Shishu Bianji Weiyuanhui (1986). *Jianguo yilai de Beijing chengshi Jianshe* (The construction of Beijing city since the establishment of the People's Republic of China). Beijing.

——(1987). *Jianguo yilai de Beijing chengshi jianshe ziliao, diyijuan, chengshi guihua* (The materials of the construction of Beijing city since the establishment of the People's Republic of China). Vol. i: urban planning. Beijing.

Beijing Review (1989). 'Hu Yaobang Mourned Nationwide', 32 (17): 5.

Beijingshi Tongji Ju (Beijing Municipal Statistics Bureau) (ed.) (1984). *Xinnxin xiangrong de Beijing* (Prosperous Beijing). Beijing: Beijing Press.

Chen, Chuankang, and Sun, Xiuping (1989). 'Beijing de fengjing luyou ziyuan' (Landscape resource for tourism in Beijing), in *Beijing luyou dili* (Geography of tourism in Beijing). Beijing: Zhongguo Luyou Press.

Chen, Zheng-xiang (1977). *Beijing.* Beijing: Guoji Yanjiu Zhongguo Wenxue Press.

Cohn, Don, and Zhang, Jingqing (1992). *Bejing Walks.* New York: Henry Holt.

Cui, Lili (1990*a*). '11th Asiad: An Unprecedented Sports Meet', *Beijing Review*, 33 (39): 16–18.

——(1990*b*). '11th Asiad Opens the Chinese Way', *Beijing Review*, 33 (40): 5–7.

Harvey, David (1989). *The Condition of Postmodernity: An Enquiry into the Origins of Cultural Change.* Cambridge, Mass.: Basil Blackwell.

He, Yeju (1985). *Kaogongji yingguo zhidu yanjiu* (Study of the building system through *kaogongji*). Beijing: Zhongguo Jianshu Gongye Press.

Hershkovitz, Linda (1993). 'Tiananmen Square and the Politics of Place', *Political Geography*, 12 (5): 395–42.

Hou, Renzhi (1979). *Lishi dilixue de lilun yu shijian* (Theory and practice of historical geography). Shanghai: Renmin Press.

——and Jin, Tao (1982). *Beijing shihua* (History of Beijing). Shanghai: Renmin Press.

Hung, Wu (1991). 'Tiananmen Square: A Polictical History of Monuments', *Representations*, 91 (35): 84–117.

Huxtable, Ada Louise (1986). *Goodbye History, Hello Hamburger: An Anthology of Architectural Delights and Disasters.* Washington: Preservation Press.

Kim, Won Bae, and Kwok, R. Yin-Wang (1991). 'Introduction: Restructuring for Foreign Investment in Asia Pacific', *Southeast Asian Journal of Social Science*, 19 (1–2): 1–13.

Kirkby, R. J. R. (1985). *Urbanization in China: Town and Country in a Developing Economy, 1949–2000 A.D.* London: Croom Helm.

Kostof, Spiro (1991). *The City Shaped: Urban Patterns and Meaning through History.* Boston: Little Brown.

Kwok, R. Yin-Wang (1981). 'Trends of Urban Planning and Development in China', in Laurence J. C. Ma and Edward W. Hanten (eds.), *Urban Development in Modern China.* Boulder, Colo.: Westview Press.

——(1996). 'A Methodological Approach to Pacific Asian Urban Forms: Development and Culture as Factors of Urbanization', *Contemporary Development Analysis*, 1 (1): 111–34.

——and Ames, Roger T. (1995). 'A Framework for Exploring the Hong Kong–Guangdong Link', in Reginald Yin-Wang Kwok and Alvin Y. So (eds.), *Hong Kong–Guangdong Link: Partnership in Flux.* Armonk, NY: M. E. Sharpe.

——and Kwok, Annette (1979). 'Le Mausolée du Président Mao', *L'Architecture d'aujourd'hui*, 201: 51–3.

Liang, Sicheng (1986). *Liang Sicheng wenji* (An anthology of Liang Sicheng), vi. Beijing: Zhongguo Jianzhuye Press.

Liu, Guang-Hua (1989). *Chinese Architecture.* London: Academy Editions.

McGee, T. G. (1991). 'The Emergence of Desakota Regions in Asia: Expanding a Hypothesis', in Norton Ginsburg, Bruce Koppel, and T. G. McGee (eds.), *The*

Extended Metropolis: Settlement Transition in Asia. Honolulu: University of Hawaii Press.

Riskin, Carl (1991). *China's Political Economy: The Quest for Development since 1949.* Oxford: Oxford University Press.

Saarinen, Eliel (1943). *The City: Its Growth, its Decay, its Future.* New York: Reinhold.

Samuels, Marwyn S., and Samuels, Carmencita M. (1989). 'Beijing and the Power of Place in Modern China', in John A. Agnew and James S. Duncan (eds.), *The Power of Place: Bringing Together Geographical and Sociological Imaginations.* Boston: Unwin Hyman.

Shoudu Shehui Jingji Fazhan Yanjiu Suo (Institute of Socialist Economic Development for the Capital) (eds.) (1989). *Shoudu fazhan zhanlue yanjiu* (Development strategy studies of the capital). Beijing: Jingji Guanli Press.

Siren, Osvald (1985). *Beijing de chengqiang he chengmen* (The walls and gates of Beijing). Trans. Yongquan Xu, ed. Tibing Song. Beijing: Yanshan Press.

Vale, Lawrence J. (1992). *Architecture, Power, and National Identity.* New Haven: Yale University Press.

Wang, Dongcan (1959). 'Beijing jianzhu shinian' (10 years of Beijing's architecture), *Jianzhu xuebao* (Architectural journal), 9 (10): 13–17.

Zhu, Zuxi (1989). 'Beijingcheng ji qiwenwu guji de zhengti baohu' (Comprehensive preservation of Beijing city and its archaeological relics), in Geng Li (ed.), *Beijing luyou fazhan zhanlue* (Tourist development strategy for Beijing). Beijing: Yanshan Press.

8
Building a Capital for Emperor and Enterprise: The Changing Urban Meaning of Central Tokyo

Takashi Machimura

Edo-Tokyo: Continuity and Discontinuity

The city of Edo (the old name of Tokyo before 1868) was first built in 1456 as a small castle by a local landlord, Ota Dokan. After a period of long warfare the Tokugawa shogunate successfully reunited the land and in 1603 established Edo as a political capital of the new regime. Thus Edo-Tokyo is a relatively new and planned city as compared to other major Japanese cities. With the establishment of the new regime Japan closed its doors to almost all foreign countries for more than two centuries. After several contacts with missions from abroad, it reopened relations with the Western countries in the 1850s, and was soon incorporated into the expanding capitalist world economy.

The modern history of Edo-Tokyo till the present day can be outlined as an intermittent process of transforming its pre-modern urban form into a modern one suitable for capitalistic development. The attempt was to make it not only a political and economic centre but also symbolic of the modernizing state. What made such a change possible? What were the spatial and cultural consequences of these historical dynamics? The purpose of this chapter is to answer these questions, focusing on the most symbolic place in Tokyo: the Imperial Palace and its vicinity, particularly the Marunouchi district.

A French critic, Roland Barthes, who once analysed the symbolic structure of Japanese society, depicted Tokyo as a city with an empty centre: 'The entire city turns around a site both forbidden and indifferent, a residence concealed beneath foliage, protected by moats, inhabited by an emperor who is never seen, which is to say, literally, by no one knows who'

I am grateful to Mary G. McDonald and Won Bae Kim for their critical and useful comments on the previous draft of this paper.

(Barthes 1982: 30). This forbidden space is still closed to ordinary people. But just next to this forbidden and invisible space, the powerful centre of Japanese economy and politics has been consistently built and rebuilt in spite of repeated change of the polity. Such a coexistence of two totally different centres in the heart of the city is one of the most important clues to understanding the cultural and historical background of modern Tokyo.

First of all, we have to consider the historical meaning of the rapid growth of Tokyo since the Meiji Restoration. Seen from a traditional viewpoint, the location of Tokyo in the Japanese regional system had been rather marginal in both geographical and cultural terms. During the long Tokugawa period, Edo as a political capital had always coexisted with two other more historic cities: Osaka as an economic centre and Kyoto as an old capital of the emperor system. Thus the historical development of Edo-Tokyo up to now means a consistent shift of political, economic, and cultural dominance from western traditional cities to an eastern developing city, Edo-Tokyo. Why and how had such a concentrated system been created? We cannot understand the historical change of modern Tokyo without answering this basic question.

The concentration of contrasting elements in a single city form has gradually created a special built environment that could serve as a physical and ideological basis of the huge capital of a premature capitalist state. The drastic change was actually impossible without an effective combination of various forces such as state power, emerging industrial capital, and social control apparatus. But, at the same time, this historical process encountered many heated disputes. The city often became a contested ground and was repeatedly reshaped.

Like other Asian cities, the mixture of indigenous non-Western (Asian) and exogenous Western (European and/or American) factors has consistently provided a distinctive framework in the urban landscape of Tokyo. But, unlike many other Asian cities, Tokyo took a unique historical course in its development, first as the capital of a rapidly modernizing imperialistic state in Asia, and later as the first non-Western critical node in the globalizing world economy. The history of this city must illustrate such a coincidence of both commonality and distinctiveness contained in a single city form.

The chapter traces this historical process briefly by dividing the whole story into four stages. The transformation of Tokyo until the 1980s directly reflected the repeated changes of the national goal for state-building: Westernization, industrialization, and economic growth. The urban history of Tokyo during this period can be also divided into three respective stages: (1) the age of 'Westernization', (2) the emergence of the modernizing metropolis, and (3) the rebuilding of the modern metropolis in the post-war period (Machimura 1994). In contrast, recent urban restructuring since the

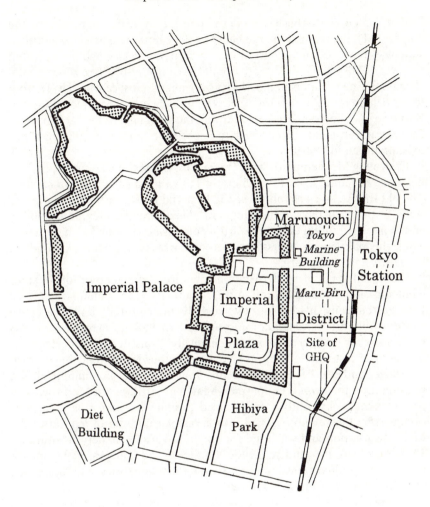

F<small>IG</small>. 8.1. Map of the Imperial Palace and its vicinity

early 1980s seems to be a quest for the post-modern form of Tokyo. The city is now entering into a new stage of history.

From a Shogunate Castle to the Emperor's Palace: Toward a Showcase of 'Westernization', 1868–1900s

In 1868 the Tokugawa regime that had dominated a country for more than two and a half centuries was overthrown by an alliance among several powerful *han* (regional domains) in the western part of Japan. New political leaders aimed to establish a modern state based on the authority of the emperor. The city which they selected as the capital of a newly established state was, however, not Kyoto where the Imperial Palace had been located for over 1,000 years, but Edo, the political centre of the Tokugawa feudal regime. For new leaders in the Meiji government, most of whom were from lower echelons of the *samurai* caste, the influence of aristocracy rooted in Kyoto seemed too outdated. In addition, the weakness of the fiscal basis of the new government forced them to choose the option of utilizing Edo's existing urban infrastructure.

The young Meiji emperor was removed from Kyoto to Tokyo, which had been just renamed from Edo. Literally Tokyo means the 'eastern capital'. The old Edo Castle was reborn as the Imperial Palace of the new emperor state. In the early days of the Meiji era, this new policy created instability and there was a threat of enemies from both inside and outside the country. Thus, this palace was, at first, guarded by barracks and parade grounds of the newly established army.

The initial stage of modernization saw direct implantation of Western elements onto a feudal setting in this Asian city. The spatial structure of Edo was basically divided into two socially and culturally unique areas: a western hilly area (*yamanote*) and an eastern flatland area (*shitamachi*) (Seidensticker 1983). *Yamanote*, the 'high city', consisted of vast estates for the *samurai* caste and numerous sites for religious institutions. In contrast, the basic land use of *shitamachi*, the 'low city', was a commercial and housing area for *chonin* or the merchants and artisans. In the process of transforming Tokyo into a modernized capital, the *yamanote* area once occupied by the *samurai* turned into a vast vacant area of land after the Meiji Restoration and was used for situating various modern institutions. These included governmental offices, schools and colleges, mansions for new leaders, military sites, modern-style national factories, and diplomatic establishments for foreign countries.

Yet this process was neither a passive inflow of Western elements nor a thoughtless imitation. Under the implicit threat of Western colonialism, the 'Westernization' policy was adopted enthusiastically for the purpose of presenting symbolic evidence of rapid modernization both for Western countries and for domestic consumption. Thus 'Westernization' during the Meiji period should be seen as an intentional and precisely endogenous process. In other words, it was one of the localized forms of Westernization in an Asian context.

This basic characteristic is relevant also to the building process of the urban landscape of modern Tokyo. Significantly, the vast land around the palace became the most important space for the Japanese version of 'Westernization'. Materializing the legitimacy, the dignity, and the progressiveness of the emerging modern state was the motive for the formation of a special built environment at that location.

First of all, the government office building district was planned as the most strategic and impressive space directly representing the political ideology of the emperor state. Several plans were presented by European architects who had been invited by the ruling political leaders. Among others, the most outstanding picture, which was drawn by two German architects, showed the idea of a complete transformation of Tokyo's urban form into a so-called neo-baroque style, not unlike Paris in the Second Empire or Berlin in the German Empire. In addition to monetary problems, internal problems within the government prevented full adoption of the plan. But some buildings such as that of the Ministry of Justice were built in the style designed by the European architects. This example, however, provides a clear idea of how the modern state is based on the emperor system in both spatial and symbolic terms (Fujimori 1982; Mikuriya 1984).

Another and more concrete evidence of localized Westernization can be found in the building process of the Marunouchi business district. In 1889 the Imperial Japanese Constitution, often known as the Meiji Constitution, was established and took effect from the following year. As the new political regime gradually stabilized, there was less need to protect the Imperial Palace with military institutions. The vast land around the palace became a wasteland for a while and was later converted to a modern space for economic activities. In 1890 the Meiji government sold the land in Marunouchi, which literally meant 'inside the castle moat', to finance the cost of removing military bases to outer areas. This vast area of land in the most crucial location was finally purchased by the Mitsubishi Company, which would later develop into one of the biggest *zaibatsu*, the huge business groups with close government connections. Mitsubishi tried to transform the vacant site of Marunouchi into a business district on the lines of a commercial street in London. Eventually, it appeared as the first modern economic centre in Tokyo. Because of its distinctive landscape in the midst of the old city of Edo, this district was often called 'Itcho London' (London Block) in the early 1900s.

Consequently, three major forces in modernizing Japan, the emperor system, the centralized state, and industrial capital, were combined both politically and spatially, and were embedded in the heart of Tokyo. As Japan took the course of a colonial empire in East Asia, this triangular power system consisting of the emperor system, the state, and capital was more firmly connected, and would lead, finally, to the destruction of not only the country but also the city itself.

Before going to the next stage, however, another and quite different example of the Westernized space built in this area deserves special mention: Hibiya Park. This was the first Western-style urban park in Tokyo. Its site had been a parade ground for the army since the early Meiji period, but by the early 1890s the Meiji government decided to build a Western-style park as evidence of its 'Westernization'. But Tokyo, which had a long history as a closed feudal city, was not familiar with the concept of a park open to the public. After a long controversy over its form, an element of the traditional Japanese garden was mixed with the idea of a Western-style park. Finally, Hibiya Park opened in 1903.

As the first public space in the midst of the imperial city, Hibiya Park often provided, unexpectedly, an important arena for various political activities and incidents. In September 1905, large crowds gathered in Hibiya Park for a mass protest movement against the terms of the peace treaty of the Russo-Japanese War. The original goals of the protest were nationalistic. But after a clash with the police during a mass meeting, the mobs attacked the buildings of a pro-government newspaper, and the police. The Hibiya Riots (Hibiya Yakiuchi Jiken) were the first urban popular protest against the government in modern Japan. Thereafter, Hibiya Park often became an important arena for political mobilization, such as a movement against a streetcar fare rise in 1906 and Rice Riots (Kome Sodo) in 1918.

No cultural and political meaning of a built environment can be realized without human interpretation and behaviour. Physical structures of the city are often planned and built exclusively by a dominating state and urban development capital, which usually seek political legitimacy and economic interests. Yet people can redefine the urban meaning more or less independently through everyday life or through collective behaviour. People always have a chance to reconstruct the meaning of various elements of the urban built environment by using them as a protest site, by using them in an unexpected way, or by ignoring them. Defining urban meaning is always a dynamic social process between structure and agency (Castells 1983; Gottdiener 1985). The modern history of Tokyo is not an exception to this argument.

Emergence of the Modern Metropolis: The Age of 'Capital of the Empire', 1910s–1945

In 1914, a new central station named 'Tokyo station' was opened just in front of the Imperial Palace. The central entrance of this station was designed for the emperor's exclusive use, but, at the same time, this station was really expected to be a main gate for the emerging Marunouchi business district. In the same year, the First World War broke out in Europe. Ironically this war rescued the Japanese economy from depression and led to an economic boom. Considerable demands for various goods stimulated further development of the manufacturing industries. Major *zaibatsu* like Mitsubishi expanded with the assistance of internal financing through the group's main bank and increasingly monopolized the market in major industrial products.

By the end of the 1920s, both industrialization and urbanization had given rise to an enormous influx of population to the Tokyo area. Particularly, as a result of both the development of manufacturing industries and the concentration of financial capital in *zaibatsu*, Tokyo's urban economy greatly expanded. The former feudal capital in the peripheral agrarian country of Japan was entering into the new stage as a huge capitalist metropolis with a population of over 5 million.

On 1 September 1923 the new building of the Imperial Hotel, designed by Frank Lloyd Wright, was to be opened for service in front of Hibiya Park. However, at noon of the same day a devastating earthquake (the Great Kanto Earthquake) suddenly hit the metropolitan area and did tremendous damage to the city. Over 100,000 died and about 550,000 buildings were destroyed. Because of the great fire caused by the earthquake, the remnants of Edo's old streets in the *shitamachi* area vanished completely.

Yet, ironically, this demolition accelerated the structural transformation of Tokyo into a modern city. While the traditional commercial centres since the Edo period such as the Nihonbashi area were heavily destroyed, the Westernized new centre for business in the Marunouchi district survived with less damage. Many newly constructed modern buildings in Marunouchi enabled a number of companies to move their offices to this expanding business district. As a result, the number of companies located in Marunouchi increased rapidly from 390 in 1921 to 1,468 in 1928 (Fuzanbo 1940). 'Itcho London' made of red-brick buildings was gradually surrounded by more functional buildings designed in the International style.

By the early 1930s, the city was almost rebuilt and appeared as a modernized metropolis. As H. D. Smith (1978: 69) pointed out, 'the city was no longer a "showcase" for novelties from abroad, but rather itself a power-house of innovation'. The change was reflected also in the class composition of the city. While self-employed merchants and craftsmen still constituted the majority in the city population, two other social groups were emerging as new classes specific to a capitalist metropolis. One was that of labourers in modern factories, the other was white-collar workers in offices. Significantly, white-collar workers in big companies and government soon became a significant force in the city. In 1928 more than 30,000 people, including 2,500 female workers, worked in the new Marunouchi business district (Chiyoda-ku 1960). These salaried workers became the core of an emerging new middle class and contributed to the diffusion of a 'modern' way of life and mass consumption culture not only within the city but also throughout the nation.

This premature modern city was soon captured by the narrow-minded nationalism and fanatic militarization of the state. Particularly the Imperial Palace and its vicinity witnessed many historical events that accelerated the political shift toward the excessively militaristic state. In 1930, because of his tolerant attitude toward naval disarmament, Prime Minister Hamaguchi

Osachi was shot by a right-wing activist in Tokyo station. The first major revolt by the nationalistic military officers and civilians took place in May 1932 (the 15 May incident). Prime Minister Inukai Tsuyoshi was killed and the Mitsubishi Bank in Marunouchi became a target of attack. In the same year, a Korean patriot who sought independence for his colonialized home country attempted to assassinate the emperor at the gate of the Imperial Palace (the Sakuradamon incident). After a short trial, he was sentenced to death and executed in the same year.

In 1936 the city centre of Tokyo once again became a site for radical nationalists to revolt. On one snowy morning in February, young military officers and their 1,400 soldiers took over the heart of Tokyo (the 26 February incident). They occupied strategic places such as the Diet building, other government buildings, and the broadcasting station. Several political leaders, including the Finance Minister, were killed. The rebels called for a drastic change of the polity ('Showa Restoration'). Three days later the coup failed and its leaders were arrested. This uprising urged the intervention of the army into politics, and later paved the way for militaristic expansion through war in China and the Pacific (Hane 1992).

In this period Tokyo was mystified as the 'capital of the emperor' or even as the 'capital of the god' by nationalistic ideologues of the militaristic state. The Imperial Palace undoubtedly stood at the centre of such a fanatic mystification. During the Sino-Japanese War from 1937 and the Pacific War, Kyujo-mae Hiroba (the Imperial Plaza) in front of the palace was often filled with huge crowds celebrating Japanese victories. The symbolic power of the palace reached its peak. But at last Tokyo, once seen as the capital of the god, was destroyed by air bombings and most buildings within the Imperial Palace were also reduced to ashes.

Rebuilding a Modern Metropolis: Prosperity and Contradictions, 1945–1970s

In August 1945 the Second World War ended. Tokyo was soon occupied by the US forces. The Allied powers' General Headquarters (GHQ), headed by General D. MacArthur, was placed in an insurance company building which commanded a view of the Imperial Palace. A lot of buildings in the Marunouchi district, which had survived the bombing with less damage due to their solid structure, were taken over by the Allied powers as commanding offices for the occupation of Japan and its post-war reforms. MacArthur acted as another emperor and American soldiers became the main figures in the heartland of the city.

The post-war development started with the rebuilding of the destroyed built environment. The drastic economic and political reforms carried out by the Allied powers undoubtedly transformed the pre-war militaristic system of Japan into a more democratic one. So far as Tokyo is concerned, however, several unchanged features between the wartime and post-war period must not be overlooked. During the war period, government control over the whole economy and society reached its peak for the purpose of mobilizing all resources and human power available for the war effort. And yet, after the war, as the democratic reform expanded its range, ironically stronger and more centralized administrative power was required for the implementation of its various policies. Therefore, in spite of several democratic reforms, political and economic power was even more highly concentrated with power holders in Tokyo. During the decades of economic growth, this centralized decision-making system was in full swing and, in effect, strengthened Tokyo's pre-eminent position in Japan's regional system in almost all fields.

On 28 April 1952, the implementation of the San Francisco Peace Treaty brought the occupation to an end. On the third day of independence, however, bloody clashes between demonstrators and police occurred in the Kokyo-mae Hiroba (the new name of the Imperial Plaza). This May Day incident became one of the most important turning-points not only in post-war Japanese politics in general but also in the history of Tokyo's urban space. During the early occupation era, the pre-war emperor system was completely demystified by MacArthur's GHQ. The Showa emperor, once regarded as a 'god' by the Japanese nation, declared openly that he was a 'human being'. Similarly, the once forbidden space of the imperial capital was also opened to the popular movements for democratization. In 1946 labour organizations and progressive leaders resumed the May Day ceremony, which had been banned by the pre-war government for a decade, at the Imperial Plaza. They often called it 'Jinmin Hiroba' (People's Plaza).

Yet by 1950, when the Korean War broke out, the attitude of the Allied powers toward left-wing social movements had changed. The GHQ banned the use of the 'People's Plaza' for the May Day ceremony in 1951. The next year the first May Day ceremony in the 'People's Palace' after independence was again banned by the Japanese government. Thus, that year labour leaders chose Hibiya Park as the approved destination of the May Day march. When the march finally arrived at the park, however, some aggressive groups headed for the 'People's Plaza' just across the street from the park. Soon serious clashes occurred between demonstrators and police. During this bloody clash, two demonstrators were killed and more than 100 were injured.

For a short period after the war, the plaza, surrounded by the Imperial Palace, the GHQ, the Marunouchi business district, and the Kasumigaseki government district, was liberated from its pre-war nationalistic image. But it was not long before this symbolic place was again caught by the persistent forces of the conservative state and the emperor system. Although the position of the emperor was redefined by the New Constitution not as a 'god' but as 'the symbol of the state and of the unity of the people', the Imperial Palace and its surrounding area still remained a mystified centre controlled by a dominating power.

By the end of the 1950s, Japan had entered the era of high-speed economic growth. In those days, while Tokyo retained its position as one of the major productive industrial centres, its economic base had already changed from manufacturing to management and services. An attempt to construct a modern industrial Tokyo, initiated during the 1910s and once interrupted by the war, reached its peak under the single-minded efforts for rapid economic growth. Particularly, for the purpose of building a highly concentrated and functional management centre for an expanding Japanese economy, a new boom of urban redevelopment started in the 1960s. Marunouchi district, which had been the most influential centre of business since the 1910s, was not an exception to such a trend.

Yet, ironically, its special location itself appeared to be an obstacle to such a redevelopment. In 1966, Tokyo Marine and Fire Insurance Company, which was one of the leading members of the Mitsubishi Business Group, decided to rebuild its headquarters in Marunouchi. According to the initial plan, its height was expected to be 127 metres (30 storeys). Until then, due to the building code for avoiding possible damage caused by earthquakes, the height of buildings had been long restricted to under 31 metres. However, as construction techniques had been considerably improved, the code had been revised in 1963. Thus, this plan would have been expected to be one of the major breakthroughs in the architectural history of Tokyo. At that time, Mitsui Real Estate, which was a member company of the rival business group Mitsui Group, had already started a project of 'cho-koso' (super high-rise) building in Kasumigaseki that was near the government building district.

Only the Marine Insurance Building, however, aroused a heated controversy over the 'aesthetic' issue and the integrity of the landscape. The reason was very simple. Since this was located just in front of the palace, special consideration on the 'aesthetic' side was required for building construction. In other words, the built environment around the palace was expected by the government to show the integrity and the dignity of the state. As Seidensticker (1990: 287) commented, traditional opinions such as 'the royal residence must not be dwarfed by huge towers' were still influential among conservative policy-makers. After a long debate, a compromise was reached, which cut five storeys from the original plan, changing its height to 99.7 metres (Seidensticker 1990; Tamura 1992). There was no convincing reason why the height should be under 100 metres. But this incident was enough to show that the district around the palace was still guided by the special forces emanating from the emperor system.

Marunouchi district still presented contested space for radicals in this period. Yet their demonstration took a totally different form. In August

1974, a powerful time bomb suddenly exploded at the gate of the headquarters building of Mitsubishi Heavy Industries Corporation in Marunouchi. It killed 8 people and injured more than 300 on the street. This was set by a Japanese left extremist group called Higashi Ajia Hannichi Buso Sensen (Anti-Japanese Armed Front of East Asia). As long as Mitsubishi's Marunouchi was the headquarters of the 'Japan Corporation', it continued to be an 'attractive' target of terrorism even after the war. This incident made it clear that this district remained a place of special political meaning.

Urban Restructuring toward a 'Global City': What has Changed and What has Not Changed in Post-'Modern' Tokyo?

In the 1980s, Tokyo entered into a new stage of development, a process that continues to the present day. The set of changes is often summarized as leading to the formation of a post-modern city and includes concepts such as post-industrialization, globalization, privatization, and the information society. A number of attractive views were introduced and combined to confirm a political legitimacy to this restructuring. Significantly, in the late 1980s, the concept of a 'global city' became a leading image of future urban growth (Machimura 1992; Douglass 1993). A temporary coalition for global city formation was organized by the national government, the Tokyo metropolitan government, urban development capital, and the ruling conservative party (Liberal Democratic Party). An urban development boom that was accelerated by the economic bubble drastically eroded the traditional spatial structure of this modern city (Machimura 1994).

During this decade, the Marunouchi district still played a crucial role in the Japanese business community. In 1980, eight of the top twenty manufacturing companies placed their headquarters in Marunouchi and two in Otemachi next to Marunouchi. In the early 1980s, Mitsubishi Estate Company gained about 6 billion yen every month from its tenants in thirty-two buildings in Marunouchi (Asahi Shimbun Shakaibu 1986). Moreover, globalization of both Japanese and foreign capital has added a relatively novel role for this district: to function as an international financial centre and a command post of global capital. As a result, various supporting institutions, such as foreign banks and security companies, became more concentrated in this small area.

This concentration, however, led to a decentralization and geographical reorganization of urban functions in Tokyo. A number of huge projects were planned and some of them were implemented. Yet only a few of these projects were in Marunouchi, where the land price was extremely high and available sites were too limited. Instead they were more concentrated in the waterfront area and other subcentral areas. Such a tendency was supported by the 'multicentric' urban structure policy adopted by the Tokyo metropolitan government (Tokyo-to 1986).

Not surprisingly, in 1988 the Mitsubishi Estate Company announced its so-called 'Marunouchi Manhattan Plan'. This plan included a daring idea of building sixty high-rise buildings of more than forty storeys in the Marunouchi district. Such a blueprint reminds observers of the Rockefeller Centre in New York that was bought by Mitsubishi Estate in 1989. Like many other plans in the era of bubble economy, this idea cannot be seen as a realistic and feasible plan. Its real purpose would be a more symbolic one. However, even in such an unrealistic scenario, one can easily detect a feeling of crisis in the Mitsubishi management, which at that time was facing fierce competition in the business office real estate market.

After the bursting of the economic bubble in the early 1990s, Japan entered a long period of recession. As the real estate market shrank, many big development plans were delayed, frozen, or often abandoned. The blueprint of the future 'global city' lost its brilliance and a daydream during the bubble age turned into a nightmare in a post-bubble stagnant economy. A final evaluation of the impact of the bubble economy on urban development is not possible now. Yet it is apparent that the first phase of urban restructuring has already come to an end.

At present the influence of the emperor system on urban formation seems more obscure than before. It is partly due to the fact that Tokyo has become so large that the sites of major projects are geographically dispersed far away from the palace as a symbolic centre. But the main reason is more structural. As post-war Japanese society has become more democratized and has experienced political conflicts between the conservatives and the left over the emperor system, the emperor himself has certainly become less significant as a source of political legitimacy.

Yet even today, the heartland of Tokyo is still not fully emancipated from the invisible power of the emperor system. This fact was suddenly made clear when the Showa emperor became seriously ill and died in 1989. During this period the city of Tokyo was forced into a state of mourning and the Imperial Palace temporarily became a strong, visible centre of the city for more than three months as programmes on television continued to disperse the live picture of the palace all over the nation. On the day of the emperor's death, which was early one winter morning, more than 129,000 people visited the Imperial Palace to express their condolences (NHK Hodokyoku 1989).

The Imperial Plaza was shrewdly transformed into a temporary stage for displaying the 'sorrow' of the nation and the 'dignity' of the system. On the day of the funeral ceremony, this stage reached its climax. In order to represent such 'sorrow' and 'dignity', the urban space was thoroughly controlled in various ways: physically by forces of police, socially by 'self-controlled' institutions such as companies and *chonaikai* (neighbourhood associations), and symbolically by the highly conforming mass media. The day after the funeral, however, the city resumed activities at full speed as if it had forgotten everything that had happened for the past several months. The Imperial Palace once again gradually became an invisible centre as it had been before the emperor's death.

As Tokyo develops globally oriented economic functions, the locally rooted culture is paradoxically beginning to gain more symbolic importance (Machimura 1994). A growing 'post-modern' type of thinking, often accompanied by a feeling of 'post-Westernization', contributes to a re-evaluation of traditional values and cultural elements. Currently there is no consensus about the city's new identity. Yet various visions for the city combine the utopian image of the good old days of Edo-Tokyo and the future image of a glorious 'global city'. Although these views seem to be contradictory, both of them are seeking a redefinition of the Westernized image of the city in a more endogenous way. Today 'Re-Japanization', or sometimes 'Re-Asianization', of the Westernized city provides another context for urban discourse on Tokyo.

Conclusion

The brief description of the urban history of the Imperial Palace area alone does not allow an illustration of the total picture of culture and built environment in modern Tokyo. Nevertheless three concluding remarks can be made at this point.

First, the entire history of urban development in modern Tokyo can be characterized as a recurrent joining of indigenous spatial and cultural background and exogenous modern elements from Western countries. Despite its lack of colonial experience, it is obvious that the basic structure of Tokyo's built environment is not an exception to other East Asian cities. The spatial coexistence of contrasting or even conflicting modes of built environment based on different cultural backgrounds is one of the most impressive features of Asian cities that are trying to catch up with Western nations through rapid modernization. 'Multiple landscapes' (Douglass, Ch. 3 above) have provided the basic context of political economy and social life in these cities.

Political and cultural consequences of such a historical experience can, however, take various forms in each society. One typical reaction to cultural and spatial chaos is a 'return' to the authentic culture and tradition (Said 1993). Recently such 'returns' have produced various forms of religious and nationalist fundamentalism in many countries. In contrast, particularly in many Asian cities, historical experience has often created a tolerant attitude toward a spatial hybridization of different cultural elements. Interestingly, such a tolerance toward multiple landscapes seemed to perform an active role in the building of the 'post-modern' city. As C. Jencks (1978) mentioned in his early polemical work, 'radical eclecticism' is one of the most basic features of post-modern architecture. This fact seems to explain partly the reason why some Asian cities such as Hong Kong or Tokyo became the most typical showcases of post-modern architecture in the 1980s.

Second and more importantly, such a 'Westernization' process was not a passive inflow of Western elements but a more interactive process based on the socio-economic and political context of each city. 'Westernization' is undoubtedly a key word for understanding the historical change of most Asian cities in both colonial and post-colonial eras, but its connotative meaning can be broader than usually expected. For instance, an attempt to transplant modern institutions into a traditional society, culture, and space was usually carried out by an authoritarian state. Yet its political power was based on very different forms of the polity, such as colonial empires, developmentalist capitalist states, or socialist states. During the pre-war period, Japan's modernization was driven by the capitalist state based on the emperor system, which was re-established by the Meiji government as an ideological means of political domination and social integration. This premature modern state tried to confirm its historical identity and political legitimacy by relating itself to the increasingly mystified emperor system. This is the main reason why the Imperial Palace area continued to hold an important political meaning through the entire period of pre-war modernization.

Even in the post-war period, in spite of the democratic reform of the polity, this position has actually remained unchanged. Although repeated changes such as industrialization, the war, and economic growth have greatly transformed Tokyo's built environment, the symbolic structure of the urban space has not experienced a major change. Undoubtedly, the Imperial Palace is still, as Barthes mentioned, an 'empty centre' of the city. Certainly the political meaning which this centre represents has been redefined frequently, yet its position as the source of symbolic power has never changed. Thus, in spite of, or rather because of, its invisible nature, the Imperial Palace has been able to remain the centre of such a political symbolism.

And yet, the historical process of the (re)formation of built environments has always been dynamic and, more importantly, conflicting. This chapter has tried to illustrate that despite the city's static and often over-mystified image, the urban history of Tokyo is not an exception to this general argument. Even the most controlled space of the city, such as the Imperial Palace and its vicinity, has witnessed consistent struggles among various actors over its definition of urban meaning. This simple but often ignored fact is the most crucial point that this chapter has attempted to clarify.

Tokyo, today, is at a historical edge. Since the end of the Edo period, Tokyo, as well as Japan, has consistently taken the advanced Western societies as a convenient model for development. For this purpose, Japan has enthusiastically tried to 'escape from Asia' and to become a member of 'Western' countries. Yet now, when most Japanese finally believe that this dream has come true, ironically, the fact that Japan still belongs to the East Asian world is again gaining recognition. While Japan has certainly become an affluent society like advanced Western countries, there still exists an irreconcilable difference in cultural values and social structure between the two. Thus currently Japan is looking for a substitute model to guide its future. And so is Tokyo. Can Japan re-enter Asia? Or does it pursue global hegemony? There are still a number of uncertainties about the future scenario of change.

Yet, as the East Asian and South-East Asian regions are growing at a rapid speed, urban commonalities, especially in the realms of the built environment and mass culture, are increasingly apparent among mega-cities in this area of the world. Today Tokyo, Seoul, Hong Kong, Singapore, and Taipei have various common features that separate them from Western cities. And the new prototype of Asian landscape is taking shape in these cities. Is the story of dual centres, such as the Imperial Palace and Maru-nouchi, relevant to the other Asian cities? This question seems to provide a new field of comparative study.

References

Asahi Shimbun Shakaibu (ed.) (1986). *Tokyo Chimei Ko(jo)* (Essay on the places in Tokyo). Tokyo: Asahi Shimbun.

Barthes, Roland (1982). *Empire of Signs*. Trans. Richard Howard. New York: Hill & Wang.

Castells, Manuel (1983). *The City and the Grassroots: A Cross-cultural Theory of Urban Social Movement*. Berkeley and Los Angeles: University of California Press.

Chiyoda-ku (ed.) (1960). *Chiyoda-ku shi (Chu)* (History of Chiyoda-ku). Tokyo: Chiyoda-ku.

Douglass, Mike (1993). 'The "New" Tokyo Story: Restructuring Space and the Struggle for Place in a World City', in Kuniko Fujita and Richard C. Hill (eds.), *Japanese Cities in the Global Economy*. Philadelphia: Temple University Press.

Fujimori, Terunobu (1982). *Meiji no Tokyo keikaku* (Tokyo planning in the Meiji period). Tokyo: Iwanami Shoten.

Fuzanbo (ed.) (1940). *Marunouchi ima to mukashi* (Marunouchi: present and past). Tokyo: Fuzanbo.

Gottdiener, Mark (1985). *The Social Production of Urban Space*. Austin: University of Texas Press.

Hane, Mikiso (1992). *Modern Japan: A Historical Survey*. 2nd edn. Boulder, Colo.: Westview Press.

Jencks, Charles (1978). *The Language of Post-modern Architecture*. London: Academy Editions.

Machimura, Takashi (1992). 'The Urban Restructuring Process in Tokyo in the 1980s: Transforming Tokyo into a World City', *International Journal of Urban and Regional Research*, 16 (1): 114–28.

—— (1994). *Sekai toshi Tokyo no kozo tenkan* (The structural change of a global city: urban restructuring in Tokyo). Tokyo: University of Tokyo Press.

Mikuriya, Takashi (1984). *Shuto keikaku no seiji* (Planning politics of capital city). Tokyo: Yamakawa Shuppansha.

NHK Hodokyoku (ed.) (1989). *Zen kiroku Showa no owatta* (The day when Showa period ended). Tokyo: Nihon Hoso Shuppan Kyokai.

Okabe, Yuzo (1993). *Rinkaibu fukutoshin kaihatsu* (Waterfront subcenter development). Tokyo: Akebi Shobo.

Said, Edward W. (1993). *Culture and Imperialism*. New York: Knopf.

Seidensticker, Edward (1983). *Low City, High City: Tokyo from Edo to the Earthquake*. New York: Alfred A. Knopf.

——(1990). *Tokyo Rising: The City since the Great Earthquake*. New York: Alfred A. Knopf.

Smith, Henry D. (1978). 'Tokyo as an Idea: An Exploration of Japanese Urban Thought until 1945', *Journal of Japanese Studies*, 4 (1): 45–80.

Tamura, Akira (1992). *Edo Tokyo machidukuri monogatari* (The story of urban development in Edo-Tokyo). Tokyo: Jiji Tsusinsha.

Tokyo-to (Tokyo Metropolitan Government) (1986). *The 2nd Long-Term Plan of Tokyo*. Tokyo: Tokyo-to.

Waley, Paul (1984). *Tokyo now & then: An Explorer's Guide*. New York: Weatherhill.

9
Hanoi: Balancing Market and Ideology

Trinh Duy Luan

Introduction

Among contemporary Asian cities, Hanoi stands out as having one of the longest histories. Any understanding of Hanoi's distinctive built environment requires the exploration of its dynamic evolution across periods marked by diverse external influences. Logan (1994) pointed out that:

the Hanoi townscape was not the product of chance forces, nor even merely the result of architectural and town planning fashions changing over time; rather each of the successive political regimes set out deliberately to impose on Hanoi its own set of beliefs about the way that urban centres should function. Each regime proceeded to design buildings, streetscapes and whole districts to demonstrate those beliefs, and, by so doing, also to demonstrate its mastery of the city and its people.

These influences stem from a long feudal period during which China played an important cultural role, a period of French colonialist rule from 1873 to 1954, and a socialist period heavily influenced by Soviet models from 1954 to 1986. During the socialist period, the Indo-China War (1945–54) and the Vietnam War (1956–75) drained Vietnam of its scarce resources which could have otherwise been used for infrastructure development. None the less, Hanoi emerged in 1976 as a rather faded but remarkably and beautifully preserved city that had been isolated from Western influences for over two decades. Throughout these periods, however, the Vietnamese have had to adapt to a series of foreign conquests and influences from China, France, and the Soviet Union. This has created a multilayered urban built environment into which the Vietnamese have inserted their own cultural influences.

Since 1986, the introduction of *Doimoi* (renovation) and the opening of Vietnam to Western influences had caused Hanoi to experience major pressures for development. This includes pressure to rebuild the urban core with 'the skyscrapers of the modern western cities', growing transportation problems, and an increase in street markets.

Today, Hanoi is a city at a crossroads. As Vietnam moves rapidly towards an open and increasingly marketized economy, Hanoi is faced with major challenges as city government and developers try to find a way of assimilating and adapting to the most recent phase of development. Thus, in contemporary Hanoi, there is acute tension between those who would like to redevelop the old parts of the city for commercial development and those who wish to preserve some of the historical layers of the whole environment. These tensions are far from being resolved.

In order to understand this situation, the chapter sets forth the major historical and contemporary cultural, economic, and political forces shaping Hanoi's built environment. The Thirty-Six Ancient Streets quarter of Hanoi, which is the traditional historical core, trading centre, and heart of the city, is selected as a case study to illustrate the interplay of these forces. Finally some observations are offered on the future of Hanoi in this phase of rapid development.

The History, Culture, and Built Environment of Hanoi

As the historical and political capital of Vietnam, Hanoi has witnessed many great historic events since its founding more than 1,000 years ago. It was a provincial capital under the occupation of a northern feudal regime in the fifth century, and from the seventh century it was an administrative centre with far-reaching control.

The history of Hanoi as a Vietnamese national capital began in the eleventh century. Legend has it that in 1010 the Ly dynasty king, Ly Thai To, defeated the northern feudal invaders, bringing an end to 1,000 years of domination. King Ly Thai then moved the capital from Hoa Lu to Dai La which is a part of present-day Hanoi. Legend states that he had a dream of a golden dragon soaring into the sky, and believing this to be a good omen, the king named the area Thang Long Thanh or 'Rising Dragon City'. The choice of the capital of Vietnam at that time was based on the traditional planning concept of 'river-in-front-and-mountain-behind'. Over the course of history, the city has had several different names: Thang Long, Dong Do, and, in 1831, Hanoi. After the end of French colonialism in Vietnam, the National Assembly of the Democratic Republic of Vietnam in July 1946 declared Hanoi the official capital of Vietnam. Thirty years later, in 1976, Hanoi became the capital of the reunified Socialist Republic of Vietnam.

Hanoi will celebrate its 1,000-year anniversary in 2010. The city has experienced at least three important historical periods during this millennium. Although Hanoi was damaged in frequent wars and conflicts, each period has left its mark on the built environment, much of which is still evident today.

The Feudal or the Thang Long Period, 1010–1883

Following the founding of Thang Long, the city was divided into two parts: the royal city, *hoang thanh*, and the commoners' city, *kinh thanh*. Thus, the architecture of the city has also had two parts: the citadel, *thanh*, and the market place, *thi*. *Thanh thi*, in Vietnamese, refers to a city or urban area, therefore itself reflecting two aspects of social and spatial division as well as the two main functions of a feudal Vietnamese city. Modelled on the Chinese city of that time, the social order of Hanoi was reflected in the spatial separation 'thanh—thi' with two main castes of mandarin and commoners. The royal city, *thanh*, belonged to mandarin and Confucian gentry or scholars who were the rearguard of mandarin while the commoners' city, *thi*, belonged to ordinary or market people.

Feudalist social hierarchy also applied strict rules and regulations on housing architecture (size, height, colour, and decoration) of various ranks of mandarin and commoners. A visible social and spatial barrier was put between them according to the Confucian principles (Hy 1993). In reality, there was some social interaction between the mandarin and commoners through cross-caste marriages. Under the domination of feudal ideology, however, the mandarin played a decisive role in shaping the social and spatial order of Thang Long-Hanoi. Furthermore, in almost all Vietnamese feudal cities, political and administrative functions were accorded primacy over economic (commercial) functions. In this case, one can say that ideology regulated the social and spatial order of Thang Long-Hanoi. During the fifteenth and sixteenth centuries, while the royal city expanded and underwent great changes under different dynasties, the commoners' city was comparatively less affected. The population in this section of the city, however, increased gradually, and its business activities became more vibrant.

With Thang Long's permanent role as the capital of the nation, royal activities became an important factor in the social life of the city during this period. This role, however, was disrupted from time to time by the invasion of northern feudal rulers. For instance, during the twenty-one-year Ming occupation between 1407 and 1428 the name of Thang Long, or Capital City, was changed to Dong Do, or Eastern City. Furthermore, due to civil wars among Vietnamese feudal dynasties, there were changes in the political role of Thang Long. However, the administrative activities in Thang Long always attracted the political forces that ruled the nation.

Despite changes in its political role during the feudal period, ancient Hanoi was always the centre of economic activities, commerce, and services. The urban economic activities were a combination of market, port, street guild, and the development of handicrafts. A distribution and marketing network was established between ancient Hanoi and the neighbouring areas. By the seventeenth century, the city was given another name, Ke

Cho, or Market Place, to reflect the importance of its economic and commercial activities. The legacy of the economic and historical character of the Thang Long culture is represented by the ancient Thirty-Six Ancient Streets quarter and guilds.

The Thirty-Six Ancient Streets quarter was first named so in 1464 when Hanoi was divided into thirty-six administrative units or guilds. The quarter was formed as early as the eleventh century and in time attracted craftsmen from the nearby rural regions. Thus under the impact of production for the needs of the nearby royal city, two aspects of community life, a community of trade and a community of residence, blended, and numerous guilds were formed as the population of the area increased. The ancient quarter has been the site for many different kinds of handicrafts such as silverwork, sculpture, leatherwork, dyeing, silk-weaving, and paper-making. Trade communities emerged clustering craftsmen from the same villages who were engaged in the same trade or handicraft. They lived in separate guilds along narrow streets that were named after the goods they sold or produced. The craftsmen from different villages also brought different religions and forms of worship to ancient Hanoi. A diversity of religious buildings were constructed: pagodas, temples, shrines, and communal houses. The pagoda was the religious place for Buddhists, the temple or shrine was the place to worship different gods, and the communal house was the place for meetings of the people engaged in the same handicraft. The communal house was also a place to worship the founder of each handicraft. The Thirty-Six Ancient Streets quarter lies north of Hoan Kiem Lake and remains the most vibrant business district in Hanoi. It is the last remaining community of ancient Thang Long, a precious cultural heritage attracting many proposals for its preservation and rehabilitation as discussed later.

In Hanoi's development, the end of the feudal period was marked by a pre-colonial period (1802–83) when the Nguyen dynasty moved the capital from Hanoi to Hue. Even when Hanoi's role as an administrative and political centre was reduced, the growth and development of economic activities continued. This economic momentum did not, however, enable Hanoi to achieve the transition to a capitalist city that occurred in other cities. The primary constraint was the strong Confucianist influence, which stressed agriculture over commerce, and the limited capability of the feudal bureaucracy (Hy 1993).

The Colonial Period, 1883–1954

In spite of its short duration, the colonial period exerted a strong influence on Hanoi's built environment because under French rule the city became the administrative centre of northern Indo-China. A mixed colonial-feudal administrative structure developed. Although social order was established by the French colonial administration, an indigenous government also contributed to this regulation along with a newly emerging Vietnamese bourgeoisie.

Two city plans were prepared during this time: Hébrard's in 1920 and Cerruti's in 1940. These plans envisaged a new centre for the city to the west, surrounded by a western boundary called Puginier Circle. Due to the complicated political circumstances and the termination of French colonial rule in 1945, neither of these plans for reconstruction of the city was implemented. While systematic widespread urban planning and management did not occur, French influences in Hanoi are quite clear. During this period, modern buildings were constructed to house commercial, service, and administrative functions. Major buildings such as the opera house, railway station, and banks were constructed in a special area of the city which became the French colonial quarter lying adjacent to the Thirty-Six Ancient Streets quarter.

The formation of the French colonial or Western quarter began in the late nineteenth century and included several Western town-planning characteristics such as a grid road and street network, wide roads and footpaths lined with stately trees, and buildings in a Western architectural style, in particular large villas with gardens. The streets were named after well-known French personalities who had made a great contribution during the colonial period in Vietnam. These included Alexandre De Rhod, Jean Dupuis, the French generals Francis Garnier and Henry Rivier, and the French governors-general Pierre Pasquier and Van Vollenhoven. The names of the Thirty-Six Ancient Streets were preserved and translated into French. For example Hang Luoc became 'Comb Street', Hang Thiec 'Tin Street', Hang Manh 'Curtain Street', Lo Ren 'Smiths Street', Hang Bac 'Silver Street', Hang Da 'Leather Street', and Hang Dao 'Silk Street'. At independence in 1945, except for the names of a few famous French scientists such as Pasteur or Yersin, all other French street names were replaced by the names of Vietnamese historical heroes.

The French quarter reflects a diversity of architectural styles. The neo-classical style is evident in Hanoi railway station, built in 1902; the French Renaissance style in the governor-general's headquarters built in 1900; the French classical style in the opera house built in 1911; the Indo-China style, with some modification to adapt to the tropical climate in Vietnam, in the Foreign Ministry building (E. Hébrard, Direction de Finances, 1931) and the History Museum (E. Hébrard, Musée L. Finot, 1925–32); and lastly the Vietnamese traditional style using traditional figures in the pavilion on Hoan Kiem Lake built in 1925.

At present, the French quarter is a multifunctional centre located in Ba Dinh and Hoan Kiem districts. The Western part of the quarter in the Ba Dinh district is now the political, diplomatic, and cultural centre of the city. The other part of the French quarter within the Hoan Kiem district has become the centre for administrative, political, economic, commercial, and service activities.

In summary, the cultural and architectural heritage of the colonial period in the French quarter is an important part of the historical built environment of Hanoi. It marks the evolution of Hanoi into a modern city. The site coverage of buildings is less than 50 per cent and overall building density is low, as houses have only one to three storeys. Low houses under the shadow of large trees is a dominant characteristic of the streetscape of the French quarter. Nevertheless, as a consequence of protracted wars and economic upheavals, the French quarter is gradually losing part of its original character. The deterioration of houses due to a long neglect of maintenance, changes in the physical structure and function of the buildings, the effects of market economic forces, and foreign investment have all had a significant impact.

The Period of Independence

After the French left in October 1954 following the Indo-China War, Hanoi became the capital of the Democratic Republic of Vietnam (North Vietnam). After ten years of post-war rehabilitation, the city was still underdeveloped and did not have a general plan for its future development. The new city government was not experienced in urban planning, social organization, and urban management. It was not until 1965 that the first preliminary master plan was prepared with the assistance of experts from the Soviet Union. But the plan was aborted because the period coincided with the American air strikes in North Vietnam. When major bombing started in 1965, the government ordered an evacuation of children and the elderly from Hanoi. Following the fiercest American bombing between 1966 and 1972, up to 75 per cent of the inner-city population was evacuated. After April 1968, although bombing was restricted to areas south of the 20th parallel, the government continued to discourage a return to the city. Administrative agencies were relocated up to 65 kilometres outside the city, factories were dismantled and relocated, and even many handicrafts industries were moved as bombing continued. The bombing destroyed or damaged many housing units leaving tens of thousands homeless, especially during the B52 operation Linebacker II in December 1972 (Thrift and Forbes 1986).

After the termination of the second Indo-China War in 1975, Hanoi gradually returned to some state of normalcy. Efforts to revive the master plan for Hanoi were stopped because of post-war economic difficulties. The centre of Hanoi including both the Thirty-Six Ancient Streets quarter and the French quarter remained essentially unchanged. Between 1981 and 1984, a new master plan for Hanoi was prepared, again with the assistance of experts from the Soviet Union. It proposed that the main axis for future development should be to the south-west. This plan, however, has not kept up to date with changing circumstances, as its implementation has been limited. In other words, up to 1985, or after more than thirty years of independence, the centre of the city was virtually the same as during the colonial period with old pagodas, French buildings, wide tree-lined streets, and many tranquil lakes.

Having undergone thirty years of socialist development, Hanoi has also become a much bigger city. As population increased the boundary of the city was extended. In 1954, Hanoi had an area of 152 square kilometres and a population of 530,000 people. In 1989, the figures was 2,123 square kilometres and 2,100,000 people respectively. New industries and living quarters were built in undeveloped areas to the south and south-west. The construction of new residential neighbourhood units followed a Soviet pattern. The new residential neighbourhood units included public services such as schools, shops, and restaurants that are surrounded by four- to five-storey apartment blocks within a radius of 500 to 800 metres from the service centre. These identical blocks created a monotonous façade. Individual flats had no private toilets or kitchens because the design was modelled after a socialist vision of a collective society. Instead, a common area for kitchens and toilets was provided at each floor. This initial model, adopted in the 1960s and 1970s, was implemented in the construction of the Kim Lien and Nguyen Cong Tru quarters. Similar blocks continued to be built in the early 1980s, with some improvements such as private toilets and some diversity in architecture, as can be seen in the Thanh Cong, Thanh Xuan, and Nghia Do quarters. In addition to these apartment blocks, many brick houses with tiled roofs were constructed. These low-rise designs supported a semi-rural life-style that led to the creation of urban villages in Hanoi.

The post-independence phase has also been characterized by several major monuments designed to symbolize the emergence of Vietnam as a new socialist nation. Typical are the Ho Chi Minh Mausoleum, the Ho Chi Minh Museum, the Workers' Palace, and the Hanoi Administrative Hall. The finance and design of these monuments were supported by the Soviet Union. During this period, the Thirty-Six Ancient Streets quarter experienced very little physical change except for the improvement of infrastructure which included the asphalt paving of the streets, and the improvement of the water supply and electrical service. Between 1954 and 1985, however, the physical condition of the buildings deteriorated significantly due to lack of restoration and maintenance. This was caused by several factors such as population pressure, overcrowding of the houses, and the overloading of the infrastructure.

The socialist model, the wartime difficulties, the highly subsidized centrally planned economic system, and the weakening of the state-owned sector of the economy have strongly affected the character of this quarter. Under the planned economy, the Thirty-Six Ancient Streets quarter changed from a traditional market place and small-scale centre of production to a residential quarter under the control of the municipal administration. Community life was consolidated and affected by the spirit of collectivization, which discouraged the development of commercial activities. The control of internal migration created a stable urban population, most of whom were wage workers employed by the state. The traditional trade activities were gradually reduced and replaced by state stores. Housing and infrastructure were neglected, the physical environment deteriorated, and the cultural and historical values associated with this quarter were severely threatened.

The built environment of Hanoi can thus be seen as a multilayered accumulation of cultural and political icons (Logan 1994). Each historical epoch and each political regime has left its influence and traces through its distinctive architecture. Under feudalism, there were the pagodas and communal houses; under colonialism, there were the smaller-scale models of French architecture; and under socialism, there were buildings reflecting the ideals of socialist collectivism in new neighbourhood units as well as political monuments. Relating the built environment of Hanoi to its Asian roots, one sees the characteristics of the urban vernacular, small low houses, narrow roads, and green-bordered urban lakes, on which have been superimposed in the past 100 years a juxtaposition of architectural styles, and an increasing diversity in the size and scale of buildings. In spite of this, the city has retained a human scale and spacing of buildings that is relevant and harmonious to people; and a landscape, streetscape, and rhythm of urban life that has been, until recently, peaceful and tranquil. These characteristics symbolize the spirit of the city that is under threat from increasingly strong market-development pressures in the 1990s.

Contemporary Forces Shaping Hanoi's Built Environment

Up to 1985 Vietnam followed a socialist development model, with a subsidized and centrally planned economy. With the adoption of *Doimoi* (renovation) at the Sixth Communist Party Congress in 1986, the country entered a new stage of development. This followed the common trend of many of the former socialist countries in Eastern Europe, the former Soviet Union, and China towards a transitional economy.

The strategy of *Doimoi* constitutes the following three main guidelines. First, a shift from a subsidized and centrally planned economy to 'a multi-sectoral and market economy with the socialist orientation and under regulation of the government'; second, democratization of social life, establishing a legitimate nation; and, third, opening up and encouragement of exchange and co-operation with the rest of the world. Implementation of *Doimoi* in recent years has achieved remarkable success in economic growth and development. Hanoi's pace of development stems from its important position in Vietnam. As the capital city, Hanoi has been the city for pilot implementation of many of the economic reforms. Being the capital city also means that Hanoi has significant access to the outside world. This 'window' has resulted in Hanoi being a major target for foreign investment. Hanoi's development is being influenced by a number of political and economic forces with different interests and tendencies to 'marketize' the economy. One result of this is the existence of twilight zones, where there is neither socialism nor a market economy based on free competition (Perkins 1994). In general, Hanoi's development is best characterized as undergoing a transition from a centralized economy to a limited market economy with associated slow changes in social, cultural, and political institutions.

The Economic and Political Changes

The two most important causal factors underlying the current transformation of Hanoi are local changes created by economic reforms, and the impact of foreign investment in Vietnam. In the first years of renovation, from 1986 to 1990, the Hanoi economy faced countless difficulties in adapting to *Doimoi*. Since then, several dynamic factors have emerged.

First, economic growth has been rapid. In 1993, Hanoi accounted for 5.6 per cent of the national GDP. The average annual rate of GDP growth in Hanoi during the three years from 1991 to 1993 was 10.5 per cent per year, reaching 13.4 per cent in 1994, compared to the national average of 7.1 per cent for the same three-year period (Vietnam GSO 1995). The total capital invested in the city enterprises comprised 11.5 per cent of the national investment.

Second, most of the growth has been in new businesses and services. According to one estimate, the efficiency of a unit of capital invested in a manufacturing enterprise in Hanoi is 1.35 while the figure for business and services is 2.48, which accounts for the great interest in these sectors (Vietnam GSO 1995). By 1993, business and services contributed 64 per cent of Hanoi's GDP.

Third, the ownership structure has drastically changed. Prior to *Doimoi*, almost all accumulated capital was incorporated in the government budget. This capital has now been decentralized to different economic sectors, especially in commerce and services. The role and power of the state-owned sector has diminished in some industries, while the co-operatively owned sector has almost disintegrated. By the end of 1993, of more than 2,000 operative business establishments in Hanoi, there were 631 limited companies, 9 shareholding companies, 107 business establishments belonging to social organizations, 340 private companies, and 928 state-run enterprises. The private-sector contribution to production has been primarily from small-scale enterprises: sub-owners, craftsmen, and the household economy.

The private sector, in various forms, has thus become an important factor in the economic life and streetscape of Hanoi. This has been facilitated in recent years by the open-door policy that permits a trade in consumer goods from neighbouring countries, especially across the northern border with China. The rise of private ownership and the effects of high growth have resulted in significant amounts of capital inflow into urban real estate, especially in Hanoi and Ho Chi Minh City. Vietnamese real estate investments are changing the appearance of the city, as mini-hotels and villas mushroom in the most desirable locations. In general, it appears that the best and most expensive locations often belong to the state, and have been reserved for joint ventures with foreigners, while the less valuable locations have become the sites for development by local investors.

Foreign investment began even before the USA lifted its embargo in April 1994. The chief sources of investment have been mainly from the Asia-Pacific region, particularly Taiwan, the Republic of Korea, Hong Kong, Singapore, Malaysia, and Japan. Foreign investors currently have a preference for hotels, services, office buildings, and some manufacturing areas such as electronics. These investors have tried to obtain prime sites in the centre of the city as well as other locations identified in the latest master plan. As may be expected with such large investments of capital, the city administration finds it difficult to reject their proposals, even going so far as to ignore building regulations and environmental impacts. Currently, there are twelve large investment projects awaiting final approval, including the twin tower ($US33.2 million), twenty-four-storey Hanoi Central Towers project on the site of the former Hanoi Hilton; and the twenty-storey Hanoi Plaza ($US41 million) project on the site of the former Hanoi General State Store near Hoan Kiem Lake. Another major project is a proposal to move nearly 1,000 households living in 150 villas in the French quarter in Hoan Kiem district, so that foreign investors can renovate and rent the villas. These projects have created considerable controversy among professionals, as well as ordinary people, who are concerned about the socio-cultural impact and the effects that the change in the built environment will have on their lives.

As may be expected, the political context of development in Hanoi also has its own unique features. In a society that was once deeply committed to politics as a top priority, the new emerging market forces and the new social groupings are challenging the existing political authority. Vietnamese society and the leadership are now rising to the challenge of building a new political order through a series of laws and regulations that have been adopted despite many difficulties and deficiencies. In this process, the mass media have more liberty in discussing a variety of issues concerning the status of citizens in a changing society, with preference for dialogues rather than one-way propaganda. The political and social atmosphere has become more open. Politics has thus taken the initial steps towards integration with economic reform.

The municipal leadership, both government and party, have acknowledged the urgent need for a transformation to a market economy, while maintaining the political stability essential for successful economic reform. In urban management Singapore is seen as a model. Whether this model can be successfully applied will be a difficult question to answer, especially considering the strict controls exercised in Singapore, as compared with the relatively unregulated situation in Hanoi. It is nevertheless clear that the Hanoi municipal leadership is open to new ideas and has explored alternatives from other South-East Asian countries.

The Social and Cultural Changes

Hanoi's social and cultural fabric has also undergone great changes under the impact of *Doimoi*. First, there has been a change in the occupational structure. Previously, 80 per cent of the labour force worked in the state-owned sector, but as a result of the recent labour flow from the state to the private sector, it is estimated that 70,000 employees, or about 20 per cent of the Hanoi labour force, have left the state-owned sector in the past four to five years. Thus, a class structure is emerging in Vietnamese society with the formation of new social groups whose incomes and living standards are directly connected to the growing market economy. Although there has not been significant privatization of the state-owned enterprises, the marketization created by *Doimoi* has resulted in a roughly even split between employment in the state and private sectors. The labour force released from the state-owned sector has found new jobs among small-scale business and service activities in the informal 'street sector' creating and adding to a new streetscape.

Second, *Doimoi* has resulted in a considerable improvement in the living standards of a majority of the residents. For example, three-quarters of respondents interviewed in a 1992 social survey reported that they had experienced either some or a considerable improvement in their living standards during the previous five years. Increased income has enabled many households to save money to improve their housing. This is a priority in view of the crowded conditions, which on average provide only 4 square metres of living space per capita. A new housing policy has also encouraged residents to rebuild their dwellings without strict controls, and this has stimulated a construction boom.

Third, there is evidence of increasing social stratification and a polarization between the rich and poor. This is a new phenomenon in Vietnam
resulting from the advantages that some social groups have had in realizing
the opportunities created by *Doimoi*, and is clearly a negative aspect of the
transition to a market economy.

Social stratification has created disparate housing and living standards
and a reallocation of space between the rich and the poor. The current
situation is not unlike the gentrification experienced in Western cities, as
the poor in the centre of Hanoi sell their houses to the rich or to joint
ventures for the construction of de luxe buildings. The rich have also built
new villas on the outskirts along the main roads that lead to the city. In the
suburban areas there are now expensive villas alongside poor farmhouses,
creating a sharp contrast in living conditions and life-style.

Previous norms and values are also changing. The concepts of the former
socialist model—equal distribution, hatred of the rich, concealment of
wealth, and derision of exploitation—are being replaced by new market-
influenced norms and values. An increasing diversity of urban life-styles is
emerging, especially the preference for a consumer life-style amongst the
rich, compared to the austere life-style of the past. It appears that the
market economy is encouraging individualism, egotism, and selfishness, and
that these norms are damaging the traditional communal unity and spirit
which has been the hallmark of Hanoi residents, most of whom were
farmers just a decade ago.

The Planning and Policy Environment: Constraints and Challenges

A well-developed and comprehensive urban planning and policy framework is critical for the effective control and regulation of new construction
and development of Hanoi. The participants in this field so far include the
municipal politicians, the policy-makers, and the professionals (planners) as
well as interventions by national leaders in some cases.

The latest master plan for Hanoi's development to 2010 was approved
by the central government in April 1992. According to this plan, the
extension of Hanoi will be mainly to the south and along some parts on the
north bank of the Red river. The Ba Dinh district will be the political
centre, and the Hoan Kiem district (including the Thirty-Six Ancient
Streets quarter) will be the commercial centre. The area surrounding West
Lake will be the cultural, recreational, and tourist centre of the city. It is
unclear, however, how much of the master plan is represented in current
construction. The implementation of the master plan will be difficult because of limited funds and lack of effective and efficient measures to control
new construction.

In Hanoi, as well as in other cities of Vietnam, an informal urban land market is developing and the illegal distribution of land is widespread. Most of the new residential construction is also illegal, in the sense that it does not have any official permission. The municipal government has set up several construction inspection teams, but their effectiveness is limited. According to one expert, up to 70 per cent of the houses which have been built in the recent past have been constructed without any permission. It is therefore apparent that there is a large gap between what the master plan dictates and what is actually happening in the construction of Hanoi.

Most Hanoi planners were trained either in Vietnam or in Eastern Europe and the former Soviet Union for conditions that are quite different from those they now experience. The new planning environment requires new planning theory, methodology, and practice, as well as advanced planning strategies, including the participation of the residents in community-based planning. The urban planners have concentrated on the design of the physical environment, using standards that may not be grounded in the present socio-economic circumstances, and with insufficient attention to the process of urban development and resource allocation under the emerging market economy. They are also severely hampered by lack of data, or by the fact that much of it is scattered amongst many different institutions. These conditions demonstrate the current difficulties and challenges for the built environment in Hanoi.

Similar difficulties confront urban policy-makers. Although the macro-policies for economic liberalization have served the spirit of renovation fairly well, many micro-level policies, laws, and regulations that have recently been adopted are probably ineffective. The legal framework is still determined primarily through the process of issuing top-down orders in the spirit of the old system. Thus bottom-up reactions and the flow of information remain weak or negligible. The higher echelons of decision-making appear to be unresponsive to the practical local-level difficulties that need to be resolved as market liberalization proceeds. In the mean time, many constraints are being created by the existing administrative system which often suffers from overlapping functions or incomplete regulatory authority.

There are also other potential development problems such as the pressure of population growth and the poor infrastructure. For example, the dangers inherent in an uncontrolled traffic situation and spontaneous construction are sending warning signs that Hanoi might become a 'second Bangkok' early in the next century. Should that occur, the unique Asian features of the city, the Thirty-Six Ancient Streets and French colonial quarters, will disappear in the swirl of polluted air and chaotic construction. A precious symbol of Vietnam's urban culture will have been lost, along with the harmony, tranquillity, and peacefulness that have been a hallmark of life in Hanoi. Within this rapid pace and haphazard pattern of development, the preservation of the historical and cultural heritage of the city has attracted great attention from Hanoi intellectuals and urban managers. Thus there are many proposals for the preservation and restoration of the Thirty-Six Ancient Streets and the French quarters, which have stimulated controversial discussion in the mass media. This suggests that there is increasing knowledge and awareness of the traditional values that have shaped Hanoi. With liberalization, public opinion has also become an important source of influence on land use.

Saving the Thirty-Six Ancient Streets Quarter

The built environment of Hanoi bears the imprint of major historical periods, reflecting both congruence and diversity in its spatial structure as well as a rich legacy of historical, political, and cultural icons. The impact of *Doimoi* has heightened the awareness of this heritage and the need to preserve and restore the city's more traditional forms.

The most controversial proposals to date have focused on measures to promote the preservation and restoration of the Thirty-Six Ancient Streets quarter, the traditional historical core of the city. UNESCO recognizes the Thirty-Six Ancient Streets quarter as a precious cultural heritage deserving preservation. The Vietnamese government has also issued regulations for the restoration and preservation of the original features of this quarter. The latest master plan 'Hanoi to 2010' confirms that the old quarter should be restored and preserved as a centre for tourism and commerce. As elaborated in this chapter, this quarter has evolved from the feudal period to become the centre of Hanoi's craft communities. As craftsmen also brought different religions and forms of worship, a diversity of religious buildings coexists in this quarter. According to a conservation study (Phe and Nishimura 1990) there are forty-five places for religious worship in this quarter. The political upheavals of the early twentieth century led to the neglect, but not the destruction, of the quarter. It was only after 1985, when Vietnam embarked on economic reforms, that the quarter was threatened by new construction projects. Today, this area covers 80 hectares in eight wards (subdistricts) of the Hoan Kiem district, and has a population of 70,000 people.

The success of *Doimoi* in the last ten years has created a conflicting set of forces which threatens the Thirty-Six Streets quarter. On the one hand, market reforms have led to a revival of this quarter as the centre of commercial activity. On the other hand, the new class of merchants who have grown wealthy from this revival want to continue doing business in this quarter, but are no longer willing to tolerate the inadequate living conditions of the tube houses populating the Thirty-Six Ancient Streets quarter. Their desire for better housing, coupled with the growth of real estate investments fed by new wealth, and a more liberal set of construction regulations, have spurred the construction of new houses, shops, and mini-hotels built with concrete, aluminium, and glass in a modern architectural vernacular. This construction has overshadowed the traditional houses and greatly affected what was a traditional streetscape marked by low, curved, tiled roofs. Thus, the recent contrast between rich and poor has become clearly evident from the juxtaposition of new and old houses along the streets.

For several years recently there have been many cries by professionals, scholars, and concerned people for the preservation and restoration of the Thirty-Six Ancient Streets quarter. The most vocal are the architects, urban planners, and historians who are very sensitive to the rapid changes in the city's built environment. Their arguments are clear: as an Asian city, Hanoi at present and in the future cannot be without its cultural-historical identity that is embedded in the architectural heritage of the city. However, any preservation programmes will depend on the will of the municipal authority. Faced with many challenges and problems of development control, and due to their limited capacity, the municipal authorities initially were not in favour of preservation.

The warnings by professionals and foreign experts, as well as the rapid disappearance of old buildings, have convinced the authorities to initiate preservation efforts and institute building controls for the Thirty-Six Streets quarter. A series of regulations and restrictions have subsequently been issued to control spontaneous new construction. For instance, Decision No. 3234/DQ titled 'Regulation on Construction Management and Preservation of Hanoi's Ancient Quarter' was promulgated by the chairman of Hanoi People's Committee in August 1993. This can be seen as the first step in developing legal controls associated with the preservation effort (Nguyen Lan 1993).

The latest document on this issue is Decision No. 70 signed by the Minister of Construction in March 1995, which approved the plan for pro-tection, preservation, and development of the Thirty-Six Streets quarter. According to this document, 'all new and renovated street-front houses in the quarter cannot exceed three storeys, while the height on non-street-front houses must be no higher than four storeys, or 16 metres'. Other exterior elements, such as tiled roofs and materials used to upgrade or decorate the façades of the old structures, must maintain the 'traditional harmony' of the quarter. The main purpose of the plan is to maintain the style, design and 'soul' of the area while accommodating modern living requirements (Thang 1995).

But it seems too late. It is estimated that by 1993 about 50 per cent of the old houses in this quarter had been destroyed and rebuilt (Grey 1993). The task of implementing strict municipal building regulations has been a diffi-cult one, even though an office for construction control has been set up by the Office of the Chief Architect of Hanoi. Most people know about the regulations, but they are too complicated, too expensive, and extremely time-consuming to follow up. As a result, people who cannot move to another area find it a cheap alternative to rebuild and renovate their houses on their own without following regulations and heeding restrictions.

Many foreign experts argue that the loss of the Thirty-Six Ancient Streets quarter will propel Hanoi towards joining many other Asian cities that have levelled their architectural history. Acknowledging this threat, there have been many projects proposed and submitted to municipal and central authorities. For example, there were projects proposed by Hanoi Architectural University, the National Institute for Urban and Rural Planning, the Cultural and Information Department of Hanoi City, and the Architectural Design Centre. Some projects are prepared with the assistance and support of international experts, such as 'Hanoi Planning and Development Control Project' funded by ADB, which tries to put forward a feasible strategy for conservation and development of Hanoi, especially for the Thirty-Six Streets quarter. However, due to the lack of funding and experience, little has been accomplished, while the development pressures of a market economy are increasing.

What then will happen to Hanoi when it enters the twenty-first century? Will it preserve its historic-cultural values, or will it follow the rapid development of many other Asian cities and face a number of unresolved socio-economic and environment problems? The role of the national and local governments is critical in the search for the best response to these provocative questions. Unfortunately, amidst the current complicated circumstances, the municipal authorities have few resources, and have to cope with many other pressing socio-economic issues. It seems to be that the main focus of the problem is the tension between, on the one hand, the tangible and powerful market economic forces, and on the other hand, the historical-cultural heritage represented in the built environment, and the values shared by the community, family, and individual. Together with the inexperience and hesitance of the municipality, these are challenges that need to be overcome. Hopefully, by the time a knowledgeable and skilful managerial echelon emerges, the urban problems which Hanoi is experiencing will still be manageable, and the built heritage exemplified by the Thirty-Six Ancient Streets quarter will still be there to be preserved. If it is not too late, then the rise of an experienced middle urban management will allow for the sustainable development of Hanoi in a balanced and ordered fashion.

Conclusion: Balancing Market and Ideology

The long history of Hanoi has demonstrated diversity of political regimes as well as historical and cultural layers in its built environment. In each period, as discussed in the chapter, the political and ideological forces often played a dominant role in regulating and 'kneading' the relevant social order and the spatial structure of the city. Historians also have noted that, throughout the history of Hanoi, the element *thi* (market place) has never been the first priority in comparison to the element *thanh* (citadel, royal city) (Tao 1992). This is true not only for the ancient, but also for the contemporary, development of Hanoi, when political ideas and administrative decrees are used to control and regulate the economy.

Until the 1980s, Hanoi did not understand the idea of a market economy. Therefore, it was a significant turning-point when the *Doimoi* policy was introduced, with the aim of initiating a market economy in Vietnam, particularly in the urban areas. The entry of an urban market-based economy has resulted in a tension between political and economic forces. Paradoxically, initiating economic reform and stimulating limited market economy without associated changes in political institutions has placed the system under increasing strain from the pressures of market forces.

In other words, competition and market forces are progressively 'wearing away' the political and administrative power that was once very strong. The case of the Thirty-Six Ancient Streets quarter reflects this current imbalance between market and ideology. The rich historical heritage which has been built from centuries of craft, commerce, and religious activities is now under the onslaught of forces unleashed by the *Doimoi* policy. The rising standard of living and the desire for better housing, Vietnamese investments in commerce and services, and joint-venture projects combined with a liberal construction policy and weak regulatory frameworks are creating a haphazard development pattern, as centuries-old buildings with rich architectural motifs are destroyed to make way for new construction.

In this current transitional period, there is an urgent need to restore the balance between market and ideology. Government administrative and regulatory agencies need to be reformed in order to combat the negative consequences of market forces, and the professionals and non-government civic groups and agencies need to campaign effectively against the destruction of Hanoi's built heritage.

References

Grey, Denis D. (1993). 'Cuu nguy cho Ha noi co' (Saving ancient Hanoi), *Architectural Review*, 6 (44): 21–2.

Hy, Nguyen Thua (1993). *Thang long-Hanoi the ky XVII, XVIII, XIX* (Thang Long: Hanoi in the XVII, XVIII, XIX centuries). Hanoi, Vietnam Historical Association: Social Sciences Publishing House.

Lan Nguyen (1993). 'Bao ton pho co Ha noi' (Preserving Hanoi's ancient quarter), *Architectural Review*, 6: 16–18.

Logan, S. William (1994). 'Hanoi Townscape: Symbolic Imagery in Vietnam's Capital', in Mark Askew and William S. Logan (eds.), *Cultural Identity and Urban Change in Southeast Asia: Interpretative Essays*. Geelong: Deakin University Press.

Perkins, Dwight H. (1994). 'Vietnam: Economic Reform in Driven of Dragons', in Dwight H. Perkins, David D. Dapice, and Jonathan H. Haughton (eds.), *Vietnam: Economic Reform in Driven of Dragons*. Hanoi: Vietnam National Politic Publishing House (Vietnamese trans.).

Phe, Hoang H., and Nishimura, Y. (1990). *The Historical Environment and Housing Conditions in the 36 Old Streets Quarter of Hanoi*. Bangkok: Asian Institute of Technology.

Tao, Van (1992). 'Do thi Viet nam: Lich su va hien tai' (Vietnamese cities: history and present), in Hoi thao Khoa hoc To chuc va Quan ly Do thi (Workshop on urban organization and management). Hanoi: Vietnamese Government Personnel and Organizational Department.

Thang, Ha (1995). 'Tougher Measures to Save the Old Hanoi', *Vietnam Investment Review*, 15 (21): 21.

Thrift, Nigel, and Forbes, Dean (1986). *The Price of War: Urbanisation in Vietnam 1954–1985*. London: Allen & Unwin.

Vietnam General Statistic Office (GSO) (1995). *Statistical Yearbook 1994*. Hanoi: Statistics Publishing House.

10

Hong Kong: From a Colony to a Model for China

Si-ming Li

Introduction

Hong Kong, until now a British crown colony, is often said to be 'a borrowed place on borrowed time' and a place where the East meets the West. The city is located on the south China coast and consists of Hong Kong Island, ceded to Britain under the Treaty of Nanjing in 1842, Kowloon Peninsula, ceded under the Convention of Beijing in 1860, and the New Territories, the portion leased to Britain for a period of ninety-nine years under a second Convention of Beijing in 1898. Together the entire Territory has a land area of 1,070 square kilometres (Fig. 10.1). Under the Joint Declaration on the Future of Hong Kong signed between the Chinese and British governments in 1984, the sovereignty of Hong Kong will revert back to China on 1 July 1997. From that date onwards, Hong Kong will become a Special Administration Region (SAR) of China with its own mini-constitution—the Basic Laws—under the principle of Hong Kong people governing Hong Kong.

More than 95 per cent of the Territory's population of 6 million consists of ethnic Chinese. A cosmopolitan outlook prevails, however, in many parts of the city, particularly in the central business districts of Central, Wanchai, and Tsim Sha Tsui. The expatriate population, which numbered more than 230,000 in 1992 (Kong 1993), comes from a variety of places including the United States, the United Kingdom, Canada, Australia, Japan, the Philippines, Thailand, Korea, India, and Pakistan. Professional and managerial workers account for some 30,000 of the expatriate population (Skeldon 1995), but the largest group of expatriates is probably the Filipina domestic women workers whose number exceeds 130,000 (Skeldon 1995). In addition

The author has benefited substantially from comments made by Dr Yok-shiu F. Lee, Department of Sociology, Chinese University of Hong Kong, and an anonymous referee on earlier drafts of the chapter. Also he would like to thank Ms Yin-Ha Tang, Department of Geography, Hong Kong Baptist University, for assisting in data collection.

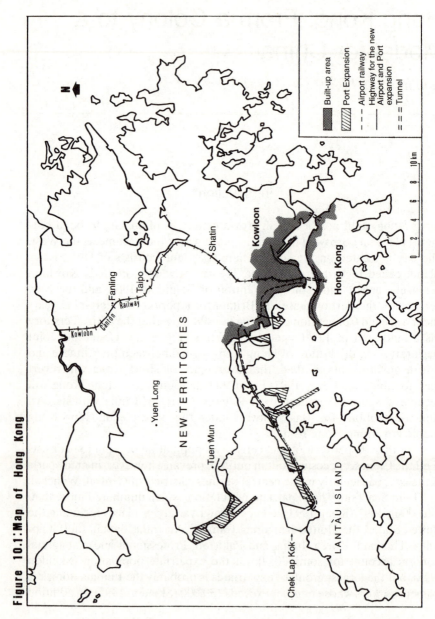

Figure 10.1: Map of Hong Kong

FIG. 10.1. Map of Hong Kong

to the resident expatriate population, more than 10 million tourists from all over the world visit Hong Kong each year. Recent developments, including China's adoption of the open policy in 1979 and the rapid growth of China's economy, and the concomitant restructuring of the Hong Kong economy which lays an increasing emphasis on trade, finance, and communications, add to the city's international flavour and point toward Hong Kong's elevation to 'world city' status (Friedmann 1986; Lo 1992).

Certainly these developments have their expression in Hong Kong's built environment. Dominating Hong Kong's ever-changing skylines on both sides of Victoria Harbour are skyscraper office towers of modern and post-modern architectural designs. Linking the various parts of the city are state-of-the-art urban freeways and railways. But against these scenes typical of a modern and thriving metropolis are images that are unmistakably Hong Kong. The bustling harbour criss-crossed daily by hundreds of ferries, lighters, tugs, junks, and sampans, and docked by an equally large number of container vessels in the 50,000-ton class; the high-rise apartment buildings closely packed together and constructed on the immensely steep slopes of Victoria Peak that seem to defy any common-sense understanding of gravity and building safety; the intensity of redevelopment activities including the tearing down of buildings that have just been completed or even are under construction, and the resultant seemingly uncompromising mix of the old and the new; the multistorey flatted factories, twenty to thirty storeys high, that are found not only in the industrial districts but in the predominantly residential areas; the public housing estates that play a leading role in new town development; the prolific Spanish-style 'village houses' that spring up in the New Territories in a city known for its land scarcity and exorbitant land prices. One can prolong this list almost indefinitely; but it is sufficiently clear that Hong Kong's built environment mirrors its cultural diversity and is the product of myriads of forces that have helped shape the city's unique path of social and economic development. In short, Hong Kong's colonial status and cosmopolitan character appear to fit nicely into what Redfield and Singer (1954) call 'heterogenetic' cities in their seminal exposition on the interrelation between culture and the city.

It is beyond doubt that economics is Hong Kong's overriding concern. In the paragraphs below, however, I shall endeavour to show that cultural and historical elements are much more than residual factors in accounting for the evolution of the Territory's landscape (McGee and Yasmeen Ch. 4 above). In particular, I shall argue that many salient features of Hong Kong's built environment, such as the Territory's high densities, the lack of social and recreation amenities and green space, especially in the older districts, the low levels of housing provision particularly in relation to the levels of income, and the apparent disregard of the adverse impacts on residential environments in the construction of transport infrastructures, while being products of specific historical circumstances, are also reflective of Hong Kong's polity and society. The characteristics of the latter, on the other hand, are largely explicable by reference to the interaction between the colonial state and the cultural traits of the local Chinese population. Of course, the built environment is not just a 'dependent variable' (Zhu and Kwok, Ch. 7 above). Different kinds of environment are conducive to different kinds of behaviour and the formation of different subgroups and subcultures. Moreover, an urban space can also be seen as a 'contested space' (Douglass, Ch. 3 above), contested not only between social groups but in the case of Hong Kong recently between two sovereign powers. A deep cultural dimension underlies much of the recent disputes between Britain and China which impinge heavily on the direction of Hong Kong's urban growth. In between the two sovereigns, there is a rising segment of Hong Kong's population which demands a greater say in the Territory's own future. In a sense the reversion to Chinese rule in 1997 will increase the orthogenetic elements of Hong Kong's culture and will give a greater significance to the symbolic meaning of the built environment. But this reversed heterogenetic transition (Kim, Ch. 2 above) definitely will not bring about a cultural landscape less heterogenetic than before.

The Legacy of British Colonial Rule

The British acquired Hong Kong to promote their trading interests in the Far East. There was no intention to build a colony in the strictest sense in that migration of British nationals to the Territory had been kept at rather low levels especially in comparison with the large numbers of Chinese moving in and out from the mainland. As Rey (1971, quoted in Douglass, Ch. 3 above) points out, dominance and transformation without dissolution of pre-colonial culture and social relations were the most important dimensions of urbanization in the colonial period in Pacific Asia. Hong Kong was no exception. Economic *laissez-faire*-ism and social non-interventionism tended to characterize British colonial rule until only recently.

The case of the New Territories

In many South-East Asian countries a model of incorporating 'pre-existing élites and political rules' (Douglass, Ch. 3 above) had been adopted during the colonial era. This model also applied to a large extent in the case of the New Territories, which constitute more than 90 per cent of the landed area of Hong Kong and which had a sizeable indigenous population prior to British occupation. The New Territories were acquired in the main to install a buffer, so to speak, between the British colony and the then politically very unstable China. In part because of this, the New Territories were administered separately from the urban areas (Hong Kong 1994). To a certain extent, the local gentry as represented by Heung Yee Kuk or the village council was given the authority to play its traditional role as the arbiter of day-to-day affairs. The New Territories Ordinance was enacted to sanction traditional practices and to protect the property rights of the indigenous population. Each male (but not female) adult of the indigenous population is still entitled to construct a three-storey village house on agricultural lands. Such village houses are often called *ding-wu* (male sibling houses) to denote the hereditary rights given to the male offspring or *nan-ding*, and are subjected only to minimal development control. Despite their feudal and anachronistic flavour, such rights were enshrined in the 1984 Sino-British Joint Declaration and subsequently the Basic Laws. Haphazard development of *ding-wu* presents a real challenge to the planner. Recently, because of the immense increase in land and housing prices, *ding-wu* construction activities have intensified and prolife-rated, and the right to construct *ding-wu* has become extremely valuable. Any attempt by the government to amend the outdated and often discri-minatory practices provided by the New Territories Ordinance, however remote from *ding-wu* construction, would be seen as an infringement on the indigenous rights and meet with strong resistance.

A recent incident relates to the implementation of the Bill of Rights. A proposal was made in 1994 to repeal the customary practice of exclusive male inheritance of properties. Heung Yee Kuk strongly opposed such a move. A protest was staged outside the Legislative Council Building on the day when the proposal was discussed in the legislature. Members of Heung Yee Kuk clashed with demonstrators in support of women's rights and resulted in a number of injuries. One could, of course, interpret this as a conflict between traditional values and Western norms, and as an idiosyn-crasy in an otherwise highly industrialized and Westernized society. But the alignment with local élites and the reliance on indigenous socio-cultural institutions in extending the British colony to the New Territories ninety years ago had already sown the seeds for the recent unrest.

Social and political non-interventionism of the colonial state

The British colonial practice of non-interventionism was extended to most aspects of social and economic affairs (Lau 1984; Lau and Kuan 1988). Until recently Hong Kong had been characterized by a clear separation between polity and society. The local Chinese people participated very little in public policy-making: representative government with a legislature consisting of elected members was practically unheard of as recently as the early 1980s. The governor of Hong Kong who was (and still is) vested with extensive constitutional powers was (and still is) appointed directly by the British crown. Top echelons of the government were almost without exception made up of expatriates coming from Britain and appointed by the governor himself. Even in middle-rank management expatriates accounted for a substantial portion of the work-force.

Immediately after the Second World War, the British had made it a policy to begin recruiting local people for more senior administrative ranks (Lau 1984). Yet localization proceeded at a snail's pace. Thirty years later in the late 1970s, at a time when Hong Kong's position as the financial centre of the Far East had been firmly established, local officers still accounted for less than half of the officers in the administrative grades (Lau 1984). Of course upward mobility within the government was not totally barred, and a system of civil servants was instituted in the main after the British meritocratic traditions. Nevertheless, it may be argued that the local–expatriate schism and the implied discrimination was not conducive to attracting the local élites to join the ranks of the government. Above all, much better opportunities can be found in the business world (Wong 1986).

Given Hong Kong's rapid economic growth and the increasing social complexity, and given the fact that an increasing number of Chinese residents have come to consider Hong Kong as their home, it has become difficult to keep the local population out of the political process. Until recently, the British colonial government had continued its effort to keep the local people out of politics by emphasizing the destabilizing effects of political activities and placing strict limits on the activities of the trade unions (Culbert 1991). But attempts had also been made, especially after the riots of 1966–7 instigated by the Cultural Revolution in China, to establish channels of communication between the British colonial government and the subjects that it ruled. These have included, *inter alia*, the establishment of consultative bodies such as the Transport Advisory Committee and enlargement of the composition of a number of policy-making bodies such as the Town Planning Board and the Housing Authority in order to draw in members of the local élites and professionals to the decision-making process.

By the 1970s and early 1980s government by consultation and consent seemed to be the norm (Lau 1984), but participation in the decision-making process was restricted to a very small segment of the population. Moreover, members of such consultative bodies were appointed by the government and did not have to be responsive to an electorate. Not only was entry to the 'debate'—the political arena and the decision-making process (Clark and Dear 1984)—under tight control, but the very nature of the debate was also strictly limited. Opposition to government policy resulted in an appointee's removal from office (Culbert 1991). Yet the introduction of a consultative process, however limited, did help to convey an impression of professionalism and impartiality in the public policy domain. Policies that were potentially unpopular—for instance the tripling of the vehicle licence fees and the first registration tax for private cars in May 1982 to control traffic growth—were justified by reference to long-term public interests. The Hong Kong government was thus able to maintain its legitimacy by portraying itself as a guardian of justice and common good.

Government by consultation implied a partial retreat from non-interventionism. Reluctantly, at first under pressure arising from exceptional circumstances, particularly the large influx of refugees from China in the early 1950s, and subsequently in response to the growing aspirations of the population reflective of the increasing affluence, the Hong Kong government began to assume a more active role in the running of the economy and in social service provision. One example of government intervention in Hong Kong is the provision of public housing. The fire in Shek Kei Mei on Christmas Day, 1954, which made 53,000 refugees from mainland China homeless had been widely considered to be the immediate cause that led to the large-scale participation of the state in the housing domain (Pryor 1983; Castells *et al.* 1990). By the early 1980s some 45 per cent of the population was residing in public housing, and this level of provision has been maintained into the 1990s.

A second and related example is the new town development programme. This programme called for concentrated development in selected sites or new towns in the New Territories. Shatin is perhaps the best known because of its proximity and good transport linkages to the urban areas. In addition there are Tuen Mun, Taipo, Fanling, Yuen Long, Tin Shui Wai, and Junk Bay. For the first time in the history of Hong Kong, urban development had spilled over the Kowloon Hills to the heart of the New Territories. Almost invariably, public housing has led this development. The public housing-led urban extension brought hundreds of thousands of mainly young and low- to middle-income households of urban origin to the New Territories. Between 1971 and 1991, the population of the New Territories increased from 676,000 to 2,052,000 (Li 1994). Not surprisingly, this development has resulted in conflicts between the newcomers and the indigenous population who retained a more rural outlook.

The scale of Hong Kong's public housing provision and the new town development programme is massive by any measure. Yet it may be argued that the launching of these programmes was dictated by circumstances. The government had tried hard to contain its level of involvement. Provision of an urban infrastructure adequate for the functioning of the economy was and still is seen to be the most important function of government. A low tax environment together with a balanced budget are among the government's overriding concerns. The former Finance Secretary Sir David Haddin Cave coined the term 'positive non-interventionism' to describe such a principle of government. Reflective of this principle, the space standards of the public housing flats and the kinds of amenities provided were and still are kept to the minimal, especially in relation to the degree of social affluence of the Territory. Also, the new towns were meant to be 'self-contained' not only in terms of provision of urban amenities but also of employment opportunities. 'Self-containment', of course, was a planning ideal. It also may have reflected the reluctance of the colonial government to commit itself in large-scale transport infrastructure investment (Lu and Gong 1985). It is interesting, perhaps, to compare the relatively low levels of infrastructure and amenities provision in Hong Kong with the case of Singapore in both the colonial and post-colonial periods (Ho, Ch. 11 below). Yet even this degree of government intervention has led some economists, especially the free market advocates (for example, Wong and Staley 1992), to proclaim that *laissez-faire*-ism in Hong Kong is a myth.

The transition to a service economy, unfortunately, has rendered self-containment an unrealizable goal. Most people who have moved to the new towns have to commute to the main urban areas to work. Invariably improvements in transport linkages lag behind traffic growth, resulting in perennial and ever-worsening congestion. Many of the new town residents, especially those in Tuen Mun and Yuen Long, have to spend four to five hours each day in commuting and therefore cannot find the time or energy to take care of their families. Their activity system is very much distorted. Many have experienced a feeling of isolation and helplessness. Ironically, one study reveals that neighbourly relations develop as a consequence so as to uphold the informal social support system. A new town subculture is evident. But whether the neighbourly relations fostered can lead to a sense of commitment to the neighbourhood remains to be seen (Tsang 1991). In the mean time, family problems and juvenile delinquency plague the new towns. Families who have moved from the main urban areas to the more remote new towns in particular exhibit immense difficulties in adjusting to their new environment.

The Hong Kong Chinese Society and the Built Environment

Materialism, utilitarianism, and the real estate market

Social and economic non-interventionism and a clear separation between the state and society have been the cornerstone of politics in Hong Kong. The fact that the Hong Kong government could cling to such a principle of government for some 150 years until only recently strongly suggests that it has a solid social base. The great majority of people in Hong Kong are Chinese. Yet Hong Kong Chinese society differs in many respects from society in traditional China and mainland Chinese society today. Lau (1984: 67) argues that the Hong Kong Chinese society represents 'an ingenious combination of typical Chinese social features and features developed in the local setting whose origin might be Chinese or foreign'. Sir David Ford (1994), former Chief Secretary for the Hong Kong government, describes the Hong Kong people as 'a rare alchemy'. Such a view is shared by other scholars such as Wong (1986).

Until as recently as the mid-1970s, the population of Hong Kong was composed largely of immigrants and their offspring. Certainly these immigrants brought with them values that could be traced back to China's 5,000 years of history. But the immigrant population was in a sense a 'self-selected' group. Immigrants in general tend to be more adventurous, and they migrate in order to search for a better life. Most Hong Kong Chinese came from Guangdong, a province distinguished by its early contacts with the West. In the case of the immigrants who fled the Chinese civil war and the communist regime to Hong Kong in the late 1940s and 1950s, many already had the experience of urban living before they came. In fact it was the immigrants from Shanghai who brought with them both the capital and skilled labour that many argued were the cornerstones of Hong Kong's industrialization. The lack of a landed gentry class except perhaps in the New Territories meant that there was an absence of powerful custodians to uphold Confucian doctrines (Lau and Kuan 1988). This, together with the 'borrowed place on borrowed time' attitude among the immigrant population, fostered materialism and utilitarianism. The quest for material advancement and hence wealth accumulation could be further attributed to the fact that 'upward mobility through political channels [was] essentially blocked and economic mobility is the only viable alternative route of gaining prestige and status' (Lau 1984: 70). Industry, enterprise, and an incessant drive to excel, not so much as a member of staff within the corporate ladder but more likely as an entrepreneur in the business world, are concomitant cultural traits (Wong 1986; Ford 1994).

Arguably, these cultural traits have been important driving forces underlying Hong Kong's economic growth. The city's success has also made it an economic haven for Chinese people from overseas. Indeed Hong Kong has served as a model of economic success. Deng Xiaoping, the *de facto* Chinese head of state, has pledged more than once to create a dozen 'Hong Kongs' in different parts of China. Yet the narrow focus on personal and family gain, as opposed to community welfare, is at least partly responsible for the frantic speculation in real estate so characteristic of Hong Kong and for the apparent lack of concern for the environment or for buildings and sites with historical or aesthetic value.

Hong Kong's export-led industrialization that began in the 1950s has been one of the major economic success stories since the Second World War. The tremendous wealth generated provided not only funds needed for continued development in the industrial sector but also huge sums of money for the development of and speculation in Hong Kong's scarce

landed properties. Rising demands combined with supplies strictly controlled by the Territory's physical geography and government's land and
housing policies almost guaranteed at the outset handsome returns to investment in the real estate sector. Indeed the most influential firms and
individuals in Hong Kong derived their wealth mostly from real estate
investment. In recent years even the small investors have been drawn by the
lucrative profits to speculate in real estate *en masse*. In most countries in
the West, home ownership represents a pride; and for most people, their
homes represent their lifelong investment. But in Hong Kong owning a
home means more than these. Capital gains derived from home ownership
especially when the leverage effects of mortgage loans are taken into account could be astronomical. Home ownership and moving from one
housing unit to another are in effect a means of speculation and wealth
generation. The home is not only a place to live but a place to make money.
It is no wonder that the 1990 Social Indicators Study carried out by the
Chinese University of Hong Kong found that 66 per cent of the respondents
aspired to owner occupation in the private sector and an additional 12
per cent to Government Home Ownership Scheme flats (Lee 1992). The
frenetic speculative activities in the real estate sector so characteristic of
Hong Kong are indicative of the materialism and utilitarianism that prevail
in Hong Kong society.

The pursuit of profit certainly provides a major impetus for urban
growth and for redeveloping the inner-city areas, thus bringing about
Hong Kong's dynamic and ever-changing skylines. But over-speculation
in the property sector often drives prices to levels beyond the reach of
all except the very rich and sows the seed for social unrest. Also the
extremely pronounced booms and busts that characterize Hong Kong's
property market could destabilize the economy and result in substantial waste in social resources. Moreover, the stress on materialism and
utilitarianism often result in a neglect of historical and aesthetic values.
Unlike the great cities of imperial China the landscape of which reflects
the Confucianist world order (Samuels and Samuels 1989; Kim, Ch. 2
above), in the case of Hong Kong the city landscape is in a large sense a
manifestation of the prevailing *laissez-faire* capitalism. This, of course,
does not mean that the built environment has not been used by the
colonial state and others to symbolize a certain world view or a certain
order. But, as a general rule, land in Hong Kong is just too precious,
and there is no room for preserving buildings, let alone districts, for historical and aesthetic interests. The magnificent Victorian-style buildings
tend to occupy large lots and central locations. For these reasons they
are among the most sought-after properties for redevelopment. Even
religious structures and schools are popular redevelopment targets and
are being torn down one after another to make way for high-rise apartment and office towers. To date, only very few buildings have been

declared objects of protection by the government under the Antiquities and Monuments Ordinance. Contests over issues such as the compaign in the early 1980s for the preservation of the Hong Kong Club Building which in many respects symbolized the colonial era seem to be the prerogatives of a few expatriates and have yet to arouse interest among the local Chinese people.

Social stability, political aloofness, and urban form

Despite Hong Kong's dynamism, a majority of the people place a high value on social stability. Stability has been particularly important for those who fled China because of civil war or the atrocities committed during the Cultural Revolution or the early days of the communist regime. Such groups tend to place very limited demands on the colonial state. In fact, they have tended to avoid political involvement (King and Lee 1984) and to value social stability above ethnic pride or apparent injustice such as the highly skewed distribution of income (Lau and Kuan 1988). Even among the younger generations who were born and brought up in Hong Kong and are thus in many respects more Westernized in their outlook social stability is generally seen as a necessary condition for economic growth. Given Hong Kong's record of incessant growth, there has been a general and consistent rise in the living standard. Many do succeed in climbing up the social ladder. Under such a circumstance, class antagonism has been kept to the minimum. Conflicts in the workplace have shown a continued downward trend in the post-war period (Lu and Gong 1985). Indeed there is a general dislike of radicalism in Hong Kong (Lau and Kuan 1988). With the rapid growth of the middle class and the increased recognition of political rights, the last twenty years have witnessed increased political activism in what has been labelled as 'the reproduction domain' (Castells 1977; Lu and Gong 1985). However, as the examples below demonstrate, it is fair to conclude that advocacy planning, as an institution, is still very much alien to Hong Kong.

The high densities that strike almost all visitors to Hong Kong are perhaps the best example to illustrate how the stress on stability and the consequential high degree of tolerance have found their expression in the urban landscape. Mention has been made of the extremely low standards of space provision and lack of amenities in Hong Kong's public housing. Outside the public housing estates the situation is perhaps even worse. In the 1960s, redevelopment projects had plot ratios (or the ratio of the floor area to the site area of a building) as high as 20:1 (Bristow 1984), with a mix of residential, industrial, and commercial uses (Sit 1986). A family of eight living in a single bed space was then a common scene. Even today, in a city known for its prosperity, some 3,000 people are living in 'cage housing', which literally is cages within a dilapidated flat (*South China Morning*

Post, 2 Dec. 1994). Schmitt (1963) was impressed in the early 1960s by the relative lack of social pathology among the Chinese families in urban Hong Kong. Subsequent studies by Mitchell (1971), Millar (1976), and Chan (1978) (see also Lee 1984) found that high-density living in Hong Kong had no or few effects on the severe forms of emotional strain and on relations within the family, and was only weakly related to biopsychic maladjustment (see also Lee 1984).

In addition to crowding, Hong Kong's population, especially those residing in low-income inner-city neighbourhoods and in the older public housing estates, tolerate a lack of public amenities and a high level of noise and air pollution due to traffic and industrial activities. Kowloon Tong, which is located about 1 kilometre to the north-west of Kai Tak international airport and directly under the flight path, is able to maintain its status as a high-class residential neighbourhood. Up till the early 1980s, the government had paid lip service to environmental concerns in infrastructure development. Needless to say, freeways cut freely through urban neighbourhoods. In some extreme cases such as the Chatham Road Flyover, passengers travelling on the elevated highway can literally shake hands with the inhabitants of the residential buildings that lie on both sides of the highway bridge. Yet scenes of residents staging protests because of this have been conspicuous by their absence.

The studies on high-density living cited above are somewhat dated. A repeat of these studies today might yield very different results. Yet, judging from the popularity of the large housing estates developed by the private developers such as Tai Koo Shing and Hwangpu Gardens in which block after block of high-rise apartments are densely packed together, giving rise to gross densities of more than 1,000 residential units per hectare, high-density living is perhaps a preferred form of residential set-up among a large segment of the Hong Kong population. An analysis of census statistics showed that, unlike cities in North America, the inner-city neighbourhoods in Hong Kong are thriving, and have been able to retain and even attract the middle- and upper-income groups over time (Li 1994). Perhaps to the surprise of outsiders, the 1990 Social Indicators Study cited above reveals that most people in Hong Kong are satisfied with their housing conditions, other than housing's exorbitant prices (Lee 1992).

Hong Kong's Changing Urban Landscape

In the previous sections the focus was on how the interaction between the colonial state and the local Chinese society effects an unique path of urban development in Hong Kong, and on the role of cultural values in this development. This section will concentrate on the more current forces underlying the development of Hong Kong's urban landscape. In particular, we shall examine how two sovereign states, Britain and China, have made use of Hong Kong's built environment to advance their respective national goals, and the part played by the local Chinese society in this development. The cultural dimension, perhaps on a different plane, once again is at play.

Symbolism and the urban built form

To the Chinese people, especially those on the mainland, the significance of the signing of the Sino-British Joint Declaration in 1984 cannot be overstated. For the first time in 150 years, China will regain a part of its territory which has been labelled the Pearl of the Orient, not by force but through negotiations on an equal basis, from a major Western power. Of course, China's victory over Japan and the recovery of Taiwan after the Second World War could be seen in this light; but then the Allied forces played the leading role. 'Countdown clocks' have been erected in the centre of Beijing and Shenzhen to commemorate Hong Kong's return to the motherland. China's long tradition of the bureaucratic state in which the centre maintains a hegemony over the individual regions suggests that Hong Kong will be subject to strong centrality forces upon reunification with China, despite the promise of a high degree of autonomy. The transition of Hong Kong to Chinese sovereignty is thus marked by conflicts, both cultural and political, between China and Britain, and between the Chinese authorities and the people of Hong Kong. These conflicts have repercussions on the direction of Hong Kong's urban growth and the form and nature of its built environment.

In Hong Kong, the Bank of China building provides a striking example of political symbolism and the growing forces of centrality emanating from Beijing. The parcel of land on which the building stands was leased to the Chinese authority in 1982 amidst the Sino-British negotiations. The building, completed in 1986, has a height of seventy-eight storeys, making it the second tallest building in Hong Kong (Central Plaza in Wanchai completed in 1991 is a few metres taller but does not command the same central location despite its name). The Bank of China building dwarfs all buildings nearby. Among these are buildings that for a long time have been equated with the colonial state: the Hong Kong Bank headquarters, Jardine's House, and, above all, the Governor's House. Many in Hong Kong believe that *fengshui* principles (Kim, Ch. 2 above) had been incorporated in the design of the Bank of China building to undermine the colonial government. This is despite the professed atheism of the Chinese Communist Party. The building has a post-modern design and is triangular in shape. From afar, the building really looks like a dagger with its edges cutting directly into the Governor's House (Fig. 10.2). Some believe that the willow

FIG. 10.2. The Bank of China building

trees at the Governor's House were planted specifically to mitigate the bank building's harmful effects.

There are other incidents in which the built environment is used to signify China's reassertion of its authority over Hong Kong and the tug of war between China and Britain. One such incident is China's insistence on the use of the Prince of Wales building, another landmark located within the HMS *Tamar* compound in the Central district, as the headquarters of the People's Liberation Army, and Britain's insistence on moving the naval base from HMS *Tamar* for commercial use with the proviso that the Prince of Wales building be retained for the People's Liberation Army (*Wah Kiu Yat Po*, 8 Sept. 1994). In May 1995 the government put tender out for sale of the first tract of land from the naval base at HMS *Tamar*. China International Trust and Investment Corporation (CITIC), the flagship state-owned business conglomerate under direct supervision of the State Council (Shen 1993), offered a record-breaking $HK3.35 billion ($US429 million) bid and acquired the 3,500-square-metre site (*South China Morning Post*, 4 Aug. 1995, B1). CITIC later announced that the site would be used for the construction of the headquarters of its Hong Kong operation. The HMS *Tamar* site is a prime site and economic considerations may be the overriding concern underlying CITIC's bid. The sequence of events leading to CITIC acquisition, however, suggest that centrality concerns may also have played a significant role. Recently, a proposal was made by certain

influential members of the SAR Preparation Committee to build a massive
cultural-cum-administrative centre on reclaimed lands right next to the
future People's Liberation Army headquarters (currently the Prince of
Wales building). Included in the plan are the future Chief Executive's
House, the Central Government Offices, a 'colonial garden' with a 'ceremonial avenue', and a massive 'cultural square' containing a monument
commemorating the return of Hong Kong to China in 1997 (*South China
Morning Post*, 25 Apr. 1996). If such a plan, or even part of it, materializes,
the landscape of central Hong Kong will undergo a major transformation.
The presence of so many icons signifying the power of the centre, accompanied by an abundance of ceremonial and 'cultural' activities demonstrating loyalty to it, will give a vivid image of what 'one country, two
systems' actually implies.

China's mistrust of Britain and Hong Kong's urban development

A deep cultural dimension underlies much of China's attitude towards
Britain and Hong Kong in this transition period. The modern history of
China was a history of repeated harassment by Western imperial powers.
This has brought about a deep mistrust on the part of the Chinese authority
toward the Western industrialized nations in general and Britain in particular. Conspiracy theories abound. There appears to be a general belief
that Britain in its final days of ruling Hong Kong will try every means to
profit from it, even at the expense of the long-term viability of the Territory.
Such a mistrust was evident in the Sino-British Joint Declaration. It was
also evident in China's dealing with the Airport and Port Development
Project and Hong Kong's democratization process.

The Joint Declaration and Hong Kong's land and housing market

In Hong Kong, all land belongs to the government in theory, and public
land is leased to individual developers and end users. Before the Joint
Declaration was signed, all privately held land in the New Territories carried a lease of 99 years, beginning from 1 July 1898. Most land on Hong
Kong Island and in Kowloon carried a 75-year lease that could be renewed
for another 75 years on payment of a premium (Li 1990). Because of its
monopoly on the supply of land, the government has a great degree of
control over land prices and the pace and direction of urban development.
To a certain extent, land scarcity is a product of government manipulation.
Apart from some brief periods, land prices in Hong Kong have been
extremely high and rising and revenues from land sales have been an
important source of income for the government (Li 1990).

   At the time of the Joint Declaration, the Chinese government appeared
to be worried that the British colonial government might sell all the land it

held and transfer the fund to Britain, leaving the future government with a precarious financial base. Accordingly, annex II was included in the Joint Declaration, restricting government land sales during the transition period from 1985 to 1997 to 50 hectares a year. Land leased to the Housing Authority for the construction of public rental housing is exempted from this limit. The appropriateness of this 50-hectare limit was to be reviewed on a regular basis by the Joint Land commission set up by the British and Chinese governments. Annex II also required that the British–Hong Kong government distribute half of its net revenue from land sales to a fund reserved for the future government. Obviously, the Joint Declaration severely restricts the outgoing British–Hong Kong government in its ability to manipulate land prices and direct urban growth. The limit on land sales and the requirement to share the proceeds with the future government have not had significant adverse effects on the present government's financial position, probably because spending has been successfully controlled. Invariably, the actual land sales agreed upon by the Joint Land Commission have exceeded 50 hectares per year, but this is still an important reference point.

The limitation has severely restricted the government's ability to dampen rising land and housing prices by selling land, and the resulting distortion in the market has been severe. Hong Kong's property market rebounded sharply after the Joint Declaration was signed. By the end of 1993, the average price of residential property had reached an all-time high of $HK50,000 ($US6,464) per square metre, or some 2.2 times the price prevailing during the previous peak in 1981 (computation based on Hong Kong Rating and Valuation Department 1990). Property prices have risen in spite of events that might be expected to depress the market, such as the heavier discounting as 1997 draws closer, the 1987 stock market crash, the 1989 Tiananmen incident in Beijing, and recent attempts by the government to contain rising prices by imposing strict limits on mortgage lending and increasing the cost of real estate transactions.

One might expect that the supply of housing in the private sector would increase in response to increased profitability. In fact, as Table 10.1 shows, private housing production peaked in 1984 and has exhibited a downward trend since then. The main reason seems to be the lack of available land for private residential development. Certainly, the private developers have had some room to manœuvre, for example, by putting parcels together through the acquisition of old buildings. The formation of the Land Development Corporation in 1987, which is a public corporation empowered with land resumption rights, has provided a further avenue for urban redevelopment (Lai 1993). Private developers can also purchase agricultural land in the New Territories and apply for development approval. Indeed, there is evidence that many developers have acquired land in this way. It has been reported that private developers were able to acquire more than 100

TABLE 10.1. Hong Kong: supply of private domestic units 1982–1993

Year	Gross supply	Net supply
1982	23,140	20,800
1983	21,620	19,430
1984	22,270	21,495
1985	29,875	29,160
1986	34,105	32,910
1987	34,375	32,770
1988	34,470	31,840
1989	36,485	33,815
1990	29,400	26,530
1991	33,380	30,205
1992	26,222	23,747
1993	27,673	25,494
1994	34,173	31,710

Note: Net supply = gross supply − Demolition.

Source: Hong Kong Rating and Valuation Department, *Property Review*, various years.

hectares in 1993 and again in 1994 (*Hong Kong Economic Journal*, 22 Mar. 1994). In most cases, rezoning approval in the New Territories is only granted for low-density development, with plot ratios of less than 1.0. Thus, the last few years have witnessed a seemingly paradoxical trend—a sudden surge of large-scale, low-density suburban developments in the New Territories at a time when land for high-rise housing was in scarce supply. For the first time in Hong Kong's history, suburban development in the Western sense is taking place on a massive scale.

The Port and Airport Development project

A second example of how mistrust between China and Britain has affected urban development in Hong Kong is the Port and Airport Development (PAD) project. It has long been recognized that Kai Tak airport, with its single runway, cannot cope with Hong Kong's increasing air traffic. The airport is considered one of the most dangerous in the world because of the surrounding topography and its proximity to high-density residential buildings. In the mid-1970s, the government decided to build a new airport at Chek Lap Kok, a small island off the northern coast of Lantau Island, and commissioned a series of feasibility studies and land-reclamation experiments. In 1983, at the height of the Sino-British negotiations, the plan to build a new airport was shelved: It was revived in 1987 when a consortium led by Hopewell and Cheung Kong Co. proposed to construct a new airport with private funds. Again, the government commissioned feasibility studies, but this time there was an attempt to link the new airport with port development in order to optimize spending on Hong Kong's transport infrastructure and to secure Hong Kong's place as a centre for international trade and information exchange, serving greater China and beyond (Taylor and Kwok 1989).

In October 1989 the government decided to launch the PAD project. The total cost of the entire project was first estimated at $HK120 billion ($US15.5 billion). It has subsequently been revised upward many times. This project will encourage development along the northern shore of Lantau Island and in the Western Harbour, thus redirecting urban growth to the western part of the Territory. The government wanted to seek private funds to help finance the scheme, but any franchise agreement would go beyond 1997, as would the repayment period for any loan. Thus the project required the approval of the Chinese government. The Chinese, however, were highly suspicious of the British interest in this project, fearing that the PAD was a plot to transfer Hong Kong's reserves to Britain and to leave the future government with a heavy financial burden.

After high-level negotiations, on 30 June 1991 the British and Chinese governments signed a Memorandum of Understanding (MOU). The British agreed to complete ten 'airport core projects' before 1 July 1997. These included, *inter alia*, the first phase of the airport itself, the North Lantau Expressway with a fixed crossing to Lantau Island, the airport railway, and the West Harbour crossing. The total loan arising from these core projects that would be outstanding at the time of sovereignty transfer should not exceed $HK5 billion ($US646 million). The British also agreed to leave $HK25 billion ($US3.23 billion) in public funds for the new government.

This, however, did not end the story. When the MOU was signed, the official cost estimate for the core projects was $HK98.6 billion ($US12.75 billion) at 1991 prices. In April 1992, the Hong Kong government revised the cost upward to $HK112.2 billion ($US14.5 billion) in 1991 prices (Chu 1992: 353). An obvious consequence was that the projected level of borrowing would exceed the $HK5 billion limit. Also there were differences of opinion as to what constituted government borrowing. The Hong Kong government proposed to set up a public corporation, the Airport Corporation, to run the future airport, and a second fully public corporation, the Mass Transit Railway Corporation, was authorized to construct and run the airport railway. The Hong Kong government argued that borrowing by these public corporations should not be counted as government borrowing, but, quite naturally, the Chinese did not agree. Furthermore, the Chinese government was unhappy that the bulk of consultancy and construction work had been granted to British firms. They noted, for example, that a British-led consortium bid of $HK7 billion ($US905 million) for the construction of the Tsing-Ma bridge, a major component of the Lantau fixed crossing, was accepted, whereas a more attractive bid of $HK4 billion ($US517 million) by a Korean firm was rejected. Protracted negotiations resumed, and an agreement was finally reached on 4 November 1994 (*Hong Kong Economic Journal*, 11 Nov. 1994). During the negotiations, the Hong Kong government did manage to go ahead with the bulk of the construction, but delays and cost escalation were unavoidable. For instance, it is now almost certain that the airport railway cannot meet the deadline for completion by 1 July 1997.

China's concern and mistrust has caused even greater delays in the port development component of the PAD. China objected to granting development rights to a consortium led by Jardine Matheson, a firm that epitomized the British empire in the Far East. This objection effectively blocked the construction of container terminal number 9 (CT9) in Kwai Chung. Lu Ping, head of the Hong Kong and Macau Affairs Office of China's State Council, demanded that the grant for developing container terminal number 9 be reopened for tender (*Hong Kong Economic Journal*, 9 Oct. 1994). Although an agreement was reached on 11 January 1996 by Britain and China that the Chinese side would endorse the outcome of negotiations between the successful consortium and the Hong Kong government over CT9 (*South China Morning Post*, 12 Jan. 1996), protracted negotiations among the container terminal operations have further delayed the construction of the terminal.

Democratization, the China factor, and urban form in the post-colonial period

China's concern and mistrust relate not only to Hong Kong's infrastructure development but also to the Territory's internal political development. To be sure, China put forward the slogan 'Hong Kong people governing Hong Kong' during negotiations with the British in the early 1980s. Perhaps reflecting a genuine concern for Hong Kong's stability and prosperity, China's communist regime has adopted a highly conservative and pro-business stance towards Hong Kong's political development. The selection of Hong Kong representatives on the Basic Laws Drafting Committee is indicative of the specific interests that China wants to preserve after 1997. 'Out of 23 representatives, 8 are wealthy businessmen, 14 come under the professional-technocratic elite category, and labour has 1 representative' (Culbert 1991: 237). A balanced budget, a low tax base, and a free market—in short *laissez-faire* capitalism—appear to be China's vision for Hong Kong in the post-colonial period. Indeed these principles are enshrined in the Basic Laws.

Subsequent developments, particularly the events of 1989, have led to confrontations between the Chinese government and Hong Kong's still small but growing pro-democracy forces. An attitude of submissiveness and aloofness from politics once characterized the Hong Kong Chinese, but beginning in the late 1970s and early 1980s, the Territory's new generation of well-educated professionals or what has been labelled the new middle class (So 1993) has increasingly demanded political participation. This group tends to have a highly Westernized outlook. Many members of this group once enthusiastically endorsed China's call for 'Hong Kong people governing Hong Kong' and supported the return of Hong Kong to the motherland (So 1993). The attitude of Hong Kong people toward China is perhaps one of love–hate (Chan 1994). The opening of China and increased interaction with the mainland have heightened the consciousness of Hong Kong's Chinese population that they are different from their *laobiao* (cousins) to the north. Although people in the Pearl river delta area of China are beginning to imitate Hong Kong Cantonese and people in Hong Kong are incorporating words from the mainland in their vocabulary, a sense of superiority prevails in Hong Kong, not only among the professionals but also among other segments of the population. Most people simply do not trust China. Many have fled: emigration today averages about 60,000 a year or 1 per cent of the total population. One consequence of this massive emigration flow is the globalization of many Hong Kong families (Kong 1993). Often it is the wife and the children who first establish residence in a foreign country, most likely Australia or Canada, leaving behind the husband to take advantage of Hong Kong's rising wages and salaries. The term 'taikongren' or 'astronaut', which rhymes with a home without a wife in Cantonese, has been coined to describe a family separated as a result of emigration. Because of the prolonged recession in the host countries in the West, quite a number of the emigrants have returned. One estimate puts the number of people in Hong Kong having the right of abode in a foreign country, including the expatriates, at 700,000 or 12 per cent of the population (Skeldon 1995). In a sense, then, Hong Kong once again becomes a sojourners' society. But unlike the sojourners in the past, the present generation of sojourners constitutes the Territory's well-to-do group.

Recent advancements in the transportation and communications technologies allow frequent movements and information exchange to sustain such a globalization process. The international experience of this segment of the population has brought home values and life-styles from the West. The evolution of Lan Kwai Fong, a back alley in the Central district of Hong Kong, into a major gathering place and entertainment venue for the Territory's expatriates and the local Chinese youths is a manifestation of this. On the eve of the 1993 New Year, tens of thousands packed the small and steep alley to celebrate the new year. The festival suddenly turned into a tragedy when twenty-one people, most of them local Chinese youths, were trampled to death (Commercial Press (HK) Ltd. 1994). The incident is a sad one, but the fact that so many local youths took part in such largely Western festivities is indicative of the change in behavioural orientation in Hong Kong. The recent suburban development in the New Territories discussed above can also be seen, in part, as a response to the change in taste and preference. Because of the exorbitant land and housing prices, high-density apartment living is still the norm.

Growing affluence, however, renders the quest for choice and excellence in terms of the external layout and internal building design of the apartment possible, at least among Hong Kong's highly paid professionals. Tennis courts and swimming pools are now standard features of residential compounds built by private developers. The more luxurious ones would have a club house containing high-class restaurants, squash courts, heated pools, and perhaps computer-simulated golf courses and skiing fields. Materialism and perhaps hedonism have long been integral features of Hong Kong society. The unsettling fate of the sojourners further exacerbates this tendency. The post-modern twist that characterizes contemporary American cities (Knox 1993) is also evident in Hong Kong.

The majority of people in Hong Kong, of course, cannot enjoy such luxuries, and do not possess the relevant credentials to qualify for emigration. They are, in a sense, captive and have to accept, however reluctantly, the inevitability of communist take-over. The sudden surge of *fengshui* mythologies (Sze 1994) and the quest for mindless laughter in popular culture (Chan 1994) help to testify to the fatalistic feeling. The restructuring of the economy, particularly the massive out-migration of the manufacturing industry, has worsened the already highly skewed distribution of income. Real wages for low-paid jobs have not risen since the late 1980s, despite consistent increase in the gross domestic product (GDP). The cage housing mentioned earlier may be an extreme. But sleeping on the pavements and underneath highway bridges are still common scenes in Hong Kong. Senior citizens without family support are particularly vulnerable. A cold spell swept Hong Kong in the Chinese New Year of 1996. Stories of old people freezing to death dominated the local newspapers' headlines. Even among the less vulnerable groups, earning a living is not easy. Securing a place in a public housing estate or a flat under the Home Ownership Scheme is among the most sought-after dreams. It is safe to say that in Hong Kong a much broader culture of survival coexists with a culture of affluence (Chan 1994).

Among the new middle class there are also those who want to stay behind. Many of them feel that a democratic government would have more legitimacy and would thus be better able to resist the interference of Beijing. Complicating the matter, of course, is the fact that Hong Kong is still under British rule. Perhaps in response to international opinion and to the growing aspirations of the local population, the British have gradually introduced representative institutions in recent years. This shift began

with the establishment of the District Boards in 1982. Subsequent developments included the introduction of elected members to the Legislative
Council, first through indirect election in 1988 and then by direct election in
1991 (Louie 1992). The 1989 Tiananmen incident in Beijing worsened the
confidence crisis, but at the same time it helped to accelerate the pace of
democratic reform.

The Chinese have been sceptical of these developments, seeing them
not only as a British plot to undermine China's sovereignty and the
future Hong Kong government, but also as part of an international agenda
to topple China's communist regime. The appointment of the present
governor, Chris Patten, led to further deterioration of relations. Patten
proposed a series of reforms to accelerate the democratization process.
These include, *inter alia*, abandoning functional constituencies for
District Board and Urban and Regional Council elections and dramatically
expanding constituencies for election to the Legislative Council. This
change effectively replaces indirect with direct elections. In 1993, negotiations broke down between China and Britain on the composition of
Hong Kong's three-tiered representative bodies—the District Boards,
the Urban and Regional Councils, and the Legislative Council. Patten
decided to go ahead with his reforms despite the repeated warnings
by China that there would be no 'through trains' in the absence of an
agreement.

Not surprisingly, Hong Kong society has become divided over this power
play between China and Britain. The younger and better educated tend to
see the position of China as reactionary, whereas the present government,
under Patten, has tried to portray itself as a progressive force. In one of the
ironies of history, the newly formed Democratic Party, generally regarded
as representing pro-democracy forces in Hong Kong, has formed a united
front with the colonial government to press for democratic reforms. China
has made public its disapproval of the Democratic Party, whose members
are, as a rule, barred from taking part in such consultative bodies as the
Hong Kong Affairs Advisers Committee and the Preliminary Working
Committee, which advise the Chinese government on matters related to the
changeover. To be fair, quite a number of Democratic Party members
have tried to dissociate themselves from what they perceive as the reactionary Beijing regime and would refuse to make contact with the Chinese
government anyway. The end result is unfortunate: the views that the
Chinese government has taken into account in formulating its policies
on Hong Kong have been quite lopsided. Moreover, the confrontational
attitude adopted by both sides is certainly not conducive to a smooth
transfer of sovereignty.

Three elections have been held since Patten initiated his reforms, one
for the District Boards, the second one for the Urban and Regional Councils, and the third for the Legislative Councils. Despite the government's

repeated efforts, only half of Hong Kong's qualified electorate has registered to vote, indicating a widespread lack of interest or confidence in the electoral process. Of the registered voters, only one-third or less actually voted in these elections. The Democratic Party and its ally, the Hong Kong Association for Democracy and People's Livelihood (HKADPL), scored important victories in all three elections.

China has made it clear that the new three-tiered representative system will be dissolved after 1 July 1997. Nevertheless, the recently formed pro-China political parties have taken an active part in the elections. Such pro-China parties include the Democratic Alliance for the Betterment of Hong Kong (DABHK), which is made up primarily of labour representatives. Pro-China newspapers and the Hong Kong Branch of the New China News Agency have openly backed such political groups. In spite of their relative inexperience, the pro-China parties have made significant inroads in the first two elections. DABHK won eight seats in the Urban and Regional Council election while the Democratic Party won 23 seats (*Yazhou Zhoukan*, 19 Mar. 1995). Such development of the democratization process sparks a note of cautious optimism, suggesting that China may work through Hong Kong's representative institutions after 1997 rather than relying on legislation by decree. Of course, China has adopted a pro-business stance and has insisted on maintaining the status quo, meaning that the future government would continue to administer Hong Kong much as the colonial government has done.

Yet times have changed. It is unlikely that the new government will be able to maintain the aura of high professionalism and impartiality enjoyed by the colonial regime, or the degree of separateness from society at large. Chances are that the new government will have to rely on the support of the population and respond to demands from different segments of society.

If this is the case, then both the Chinese government, including its representatives in the Territory, and the pro-democracy forces in Hong Kong will have to learn to accommodate each other in the post-colonial era. Competition between the two groups will take place largely within the confines of Hong Kong's representative institutions. Policies with a direct bearing on urban form—such as housing, transport, land use, provision of recreational space, and environmental conservation—may provide arenas for conflict between the pro-democracy forces and the government. It will be interesting to see what role the other social and political groups will play in this process, particularly the pro-China elements. DABHK, the largest pro-China party, has taken a similar stance on social policies to the Democratic Party and HKADPL, and it may be difficult for the future government to ignore the demands of such a broad-based coalition. Yet the trend toward greater participation will run counter to the *laissez-faire* approach, and calls for government activity could jeopardize Hong

Kong's low taxation system. One likely development is that the central government in Beijing as well as the new Hong Kong government may be forced to play a game of checks and balances, lending some support to the conservative and pro-business groups in future election campaigns.

Accompanying the 1997 changeover is a reshuffling of the Territory's population. In addition to the massive out-migration to Canada, Australia, the United States, and New Zealand over the past few years, there has also been increasing in-migration from mainland China. A large proportion of the new immigrants have relatives in Hong Kong, but they also include professionals and government officials who come to take advantage of Hong Kong's booming trade with China and to prepare for the transition. This new wave of immigrants will put heavy pressure on the Territory's housing, education, and transport systems. In the past there was a concern to assimilate new immigrants into Hong Kong society, but now, given the reversion of Hong Kong to Chinese rule, a large influx of immigrants from China could mean the assimilation of Hong Kong into the society of mainland China.

Such an assimilation of immigrants from China would certainly facilitate the Chinese government's assertion of authority over Hong Kong, but the strength of Beijing's position will also be influenced by international capitalism and by democratic developments within Hong Kong. If democracy proves to be feasible in Hong Kong, then pressure will probably build up for democratic reforms within China. Thus Hong Kong, in addition to serving as an economic model, may very well play a major role in China's social and political development in the twenty-first century. The future may be uncertain, but one thing is sure: the reversion of sovereignty to China will not diminish the degree of heterogeneity in Hong Kong's culture and urban landscape.

References

Bristow, Roger (1984). *Land Use Planning in Hong Kong*. Hong Kong: Oxford University Press.
Castells, Manuel (1977). *The Urban Question: A Marxist Approach*. London: Edward Arnold.
——Goh, L., and Kwok, R. Y. W. (1990). *The Shek Kip Mei Syndrome: Economic Development and Public Housing in Hong Kong and Singapore*. London: Pion.
Chan, Hoi-man (1994). 'Culture and Identity', in Donald H. McMillen and Si-wai Man (eds.), *The Other Hong Kong Report 1994*. Hong Kong: Chinese University Press.

Chan, Ying-Keung (1978). *Life Satisfaction in Crowded Urban Environment.* Social Research Centre Occasional Paper No. 75. Chinese University of Hong Kong.

Chu, David K. Y. (1992). 'Transportation', in Joseph Y. S. Cheng and Paul C. K. Kwong (eds.), *The Other Hong Kong Report 1992.* Hong Kong: Chinese University Press.

Clark, Gordon L., and Dear, Michael (1984). *State Apparatus.* London: Allen & Unwin.

Commercial Press (HK) Ltd. (1994). *Tulu xiangang da qushi 1994* (Major trends in Hong Kong in pictures 1994). Hong Kong: Commercial Press (HK).

Culbert, Alexander R. (1991). 'A Fistful of Dollars: Legitimation, Production, and Debate in Hong Kong', *International Journal of Urban and Regional Research*, 15 (2): 234–9.

Ford, Sir David (1994). 'A Rare Alchemy', *in Hong Kong 1994.* Hong Kong: Government Printer.

Friedmann, John (1986). 'The World City Hypothesis', *Development and Change*, 17: 69–83.

Hong Kong (1994). *Hong Kong 1994.* Hong Kong: Government Printer.

Hong Kong Rating and Valuation Department (1990). *Property Review 1990.* Hong Kong: Government Printer.

——(1994). *Property Review 1994.* Hong Kong: Government Printer.

King, Ambrose Y. C., and Lee, Rance P. L. (eds.) (1984). *Social Life and Development in Hong Kong.* Hong Kong: Chinese University Press.

Knox, Paul L. (1993). 'Capital, Material Culture and Socio-spatial Differentiation', in Paul L. Knox (ed.), *The Restless Urban Landscape.* Englewood Cliffs, NJ: Prentice Hall.

Kong, Paul C. K. (1993). 'Internationalization of Population and Globalization of Families', in Po-King Choi and Lok-Sang Ho (eds.), *The Other Hong Kong Report 1993.* Hong Kong: Chinese University Press.

Lai, Wai-Chung (1993). 'Urban Renewal and the Land Development Corporation of Hong Kong', in Po-King Choi and Lok-Sang Ho (eds.), *The Other Hong Kong Report 1993.* Hong Kong: Chinese University Press.

Lau, Siu-Kai (1984). *Society and Politics in Hong Kong.* Hong Kong: Chinese University Press.

——and Kuan, Hsin Chi (1988). *The Ethos of the Hong Kong Chinese.* Hong Kong: Chinese University Press.

Lee, James (1992). 'Housing and Social Development', in Siu-Kai Lau, Ming-Kwan Lee, Po-San Wong, and Siu-Lun Wong (eds.), *Indicators of Social Development.* Hong Kong: Hong Kong Institute of Asia-Pacific Studies, Chinese University of Hong Kong.

Lee, Rance P. L. (1984). 'High-Density Effects in Urban Areas: What do we Know and What should we Do?', in Ambrose Y. C. King and Rance P. Lee (eds.), *Social Life and Development in Hong Kong.* Hong Kong: Chinese University Press.

Li, Si-ming (1990). 'The Sino-British Joint Declaration, 1997, and the Land Market of Hong Kong'. *Review of Urban and Regional Development Studies*, 2 (2): 84–101.

——(1994). 'The Changing Spatial Distribution of Hong Kong's Population: An Analysis of the 1991 Population Census', *Asian Geographer*, 13 (1): 1–16.

Lo, C. P. (1992). *Hong Kong*. London: Belhaven.

Louie, Kin-Shuen (1992). 'Politicians, Political Parties and the Legislative Council', in Joseph Y. S. Cheng and Paul C. K. Kwong (eds.), *The Other Hong Kong Report 1992*. Hong Kong: Chinese University Press.

Lu, Da-Le, and Gong, Qi-Sing (1985). *Chengshi Zhongheng* (City profiles). Hong Kong: Huafeng Shuju.

Millar, Sheelagh E. (1976). 'Health and Well-Being in Relation to High-Density Living in Hong Kong'. Ph.D. dissertation, Department of Sociology, Australian National University, Canberra.

Mitchell, Robert E. (1971). 'Some Social Implications of High Density Housing', *American Sociological Review*, 36: 18–29.

Pryor, E. G. (1983). *Housing in Hong Kong*. 2nd edn. Hong Kong: Oxford University Press.

Redfield, R., and Singer, M. (1954). 'The Cultural Role of Cities', *Economic Development and Cultural Change*, 3: 53–73.

Rey, Pierre-Philippe (1971). *Colonialisme, néo-colonialisme et transition au capitalisme*. Paris: Maspéro.

Samuels, Marwyn S., and Samuels, Carmencita M. (1989). 'Beijing and the Power of Place in Modern China', in John A. Agnew and James Duncan (eds.), *The Power of Place*. Boston: Unwin Hyman.

Schiffer, Jonathan (1991). 'State Policy and Economic Growth: A Note on the Hong Kong Model', *International Journal of Urban and Regional Research*, 15 (2): 180–96.

Schmitt, Robert C. (1963). 'Implications of Density in Hong Kong', *Journal of the American Institute of Planners*, 29: 210–17.

Shen, George (1993). 'China's Investment in Hong Kong', in Po-king Choi and Lok-sang Ho (eds), *The Other Hong Kong Report 1993*. Hong Kong: Chinese University Press.

Sit, Victor F. S. (1986). 'Industry in a Limited Space', in T. N. Chiu and C. L. So (eds.), *A Geography of Hong Kong*. Hong Kong: Oxford University Press.

Skeldon, Ronald (1995). 'Immigration and Population Issue', in Stephen Y. L. Cheung and Stephen M. H. Sze (eds.), *The Other Hong Kong Report 1995*. Hong Kong: Chinese University Press.

So, Alvin Y. (1993). 'Hong Kong People Ruling Hong Kong! The Rise of the New Middle Class in Negotiation Politics', *Asian Affairs: An American Review*, 20 (2): 67–87.

Sze, Stephen M. H. (1994). 'Dazhong quanbo' (Mass media), in Commercial Press (HK) Ltd., *Tulu xiangang da qushi 1994* (Major trends in Hong Kong in pictures 1994). Hong Kong: Commercial Press (HK).

Taylor, Bruce, and Kwok, R. Yin-Wang (1989). 'From Export Center to World City: Planning for the Transformation of Hong Kong', *Journal of the American Planning Association*, 55 (3): 309–22.

Tong, Irene (1993). 'Women', in Donald H. McMillen and Si-wai Man (eds.), *The Other Hong Kong Report 1994*. Hong Kong: Chinese University Press.

Tsang, Yuk-Ying (1991). 'Residential Satisfaction and Adjustment in a New Town: The Case Study of Public Housing Households in Yuen Long. BA thesis, Department of Geography, Hong Kong Baptist College, Hong Kong.

Williams, R. (1982). *The Sociology of Culture*. New York: Schocken Books.

Wong, Richard Y. C., and Staley, Samuel (1992). 'Housing and Land', in Joseph Y. S. Cheng and Paul C. K. Kwong (eds.), *The Other Hong Kong Report 1992.* Hong Kong: Chinese University Press.

Wong, Siu Lun (1986). 'Modernization and Chinese Culture in Hong Kong', *China Quarterly*, 106: 306–25.

11

From Port City to City-State: Forces Shaping Singapore's Built Environment

Kong Chong Ho

The built environment of the city at any point in time represents a confluence of various forces. While the formation and reproduction of cities depend on the mobilization, extraction, and concentration of surplus (Harvey 1973), this economic imperative is being mediated by a political process in which the state takes an active role in ensuring the reproduction process through intervention in production and consumption (Castells 1981). The structural approaches (see Castells 1981; Peterson 1981) to state planning have assumed a stable social and economic environment which allows planning and state action to be theorized, but actual case studies have indicated that the situation is otherwise. For example, Eckstein's (1990) study of El Centro, an inner-city settlement in Mexico city, indicated the importance of: (1) historical factors behind current social relations; (2) shifting issue-based alliances between various groups; and (3) even incidental events such as the World Cup championship soccer matches that took place in Mexico City. Thus, apart from the routine housekeeping functions, all government efforts to manage and shape the built environment are open to various pressures and interpretations.

In this complexity, culture plays a vital role. Duncan and Ley (1993) stress that culture should not be treated as a residual factor in attempting to understand the built environment. Cultural forces manifest themselves through the symbolism of built forms; they are also represented wherever values, traditions, status differences, and the general mode of social organization are embodied in the urban environment (Lawrence and Low 1990). Cultural values and traditions are also used to legitimize the actions of the state. When government policies are contested, the process of legitimiza-

I am grateful to Dr Hua-Chew Ho and Associate Professor Yeow-Chin Wee for assisting me in my research on nature conservation in Singapore

tion is especially important, and alternative cultural values may be used to advocate alternative programmes.

Singapore provides a case study of competing cultural forces and shows how a dominant value system (summarized as developmentalism) is open to challenge. The city's developmentalist orientation evolved under two political regimes. The British colonial government focused narrowly on a set of commercial interests, ignored social programmes, and created a dualistic structure in which an excellent port coexisted with overcrowded slums. The post-colonial government, headed by the People's Action Party (PAP), maintained the primacy of economic interests but broadened the economic base and included social programmes such as housing, education, and health. Broader government intervention in the post-colonial era resulted in a built environment that was increasingly controlled by the state.

The last ten years have seen the change in value system guiding Singapore's government and society. In the late 1980s, the government discovered that environmental improvement was a useful way to create a positive image at home and abroad. Thus, new symbols of governmental achievement include clean rivers and nature reserves. Such government programmes have also changed public perceptions of the environment. The golf course incident described later in this chapter illustrates a situation of tension between two competing value systems, developmentalism versus nature conservation.

The Colonial Legacy

The economy

The establishment of Singapore as a trading port by the British in 1819 was an outcome of several factors. The increased trade between India and China in tea, cotton, and opium meant heavier traffic by British ships through the South-East Asian Archipelago. The British also became gradually more involved in the region itself, particularly in the lucrative spice trade, and with this involvement came a greater concern to counter Dutch monopolistic practices in the region. Finally, the British realized that Penang, an existing port about 885 kilometres north of Singapore, was not well sited to take advantage of these trade developments (Wong 1960; Webster 1987). Under the British, practical considerations of trade and commerce never wavered, and these concerns left an indelible imprint on Singapore's landscape.

As a port city in the British colonial empire, Singapore served a function similar to that of the older coastal settlements that proliferated along the trade routes of South-East Asia. As a product of colonization, however, and as a city of commerce built from scratch, Singapore has always played a particularly strong role as a centre for the diffusion of new ideas. In this sense, Singapore was very much a heterogenetic centre (Redfield and Singer 1954), in marked contrast to the older capital cities of the region that functioned as orthogenetic centres perpetuating the cultural heritage of the nations. As an alien creation and without a countryside to provide a sense of history and culture, Singapore existed as a modern heterogenetic anomaly in South-East Asia, a fact that O'Connor (1983) acknowledged but did not conclusively resolve.

The overwhelming emphasis on trade and Singapore's ensuing prosperity led to the growth of an infrastructure of supporting industries: finance, communication, transportation, and fuel. Thus, Singapore become, as a result of its entrepôt economy, a regional centre of finance and communication, a fuel depot, and a transportation hub.

Social issues

In managing Singapore, the colonial government pursued a narrow economic policy that focused exclusively on trade. There was little indigenous manufacturing, partly because the colonial government discouraged industrial development (Ho 1993*a*) through an unfavourable tariff structure (Dixon 1991) and through other means. The colonial government had placed economic priorities above social concerns. The maintenance of the physical infrastructure of trade was of paramount importance in ensuring surplus that accrued to British interests. This imbalance of priorities is aptly summarized by Edwin Lee: 'The British in Singapore measured success by the amount of trade secured and the miles of roads paved. But what about schools, hospitals, and low-cost housing? In the matter of schools and hospitals, the government invoked laissez-faire principles and expected philanthropic contributions from private individuals who had benefited from the free trade system' (1989: 40). It is important to point out that this disregard of social and community services was only possible because the government was kept in place by effective threat of force rather than through the popular will of the people. In such a situation, the concerns of society and state only overlapped with regard to a narrow set of economic issues. Social improvement schemes were left to the various local communities. Such a system in turn meant that popular support tended to be mobilized and therefore confined within respective communities, and it never expanded past these boundaries.

While the communities did achieve some success, particularly in the area of education, their efforts tended to be limited and sporadic because they depended on the resources and energies of a few prominent individuals. Such a system did not adequately meet the communities' needs in areas such as health and housing, which required more resources and systematic planning. The combination of government inattention and limited mobilization of local communities meant that rapid urbanization of Singapore resulted in a central area that was highly congested and lacking in adequate sanitation.

After the Second World War, the colonial government showed an increasing commitment to social issues, as evidenced by the modest housing construction initiated by the Singapore Improvement Trust, Singapore's first urban improvement and housing authority (see Fraser 1949; Ho 1993*b*). The government began to recognize Singapore's over-dependence on trade and acknowledged a need to diversify into the manufacturing sector (Allen 1950; Fraser 1955). There was also a willingness to prepare for the gradual transfer of power to an independent government, while at the same time attempting to safeguard British interests (Stockwell 1986). But this awakening was only in the infant stages of realization when in 1959 Singapore became a self-governing nation-state.

The built environment of colonial Singapore

By the early 1900s, the spatial character of Singapore's commercial districts was already well entrenched. The financial district was located close to Singapore river, the site of the earliest maritime and port activities. After the Tanjong Pagar Docks were built in the 1860s, the city expanded westward along the coast. Around the turn of the century, Pulau Bukom and other islands were developed for fuel storage and bunkering. This economic infrastructure was matched by an administrative centre on the eastern side of the Singapore river, protected by a ring of military installations that, by the 1950s, occupied one-tenth of the colony's land area (Humphrey 1985).

The city centre featured a range of social amenities for the rich. Large department stores selling British and other imported goods were located along High Street and Raffles Place, together with jewellery and textile stores, tailoring shops, and cafés (URA 1989; Bellet 1969). This area also contained several symbols of the British colonial empire: the Town Hall, the Victoria Memorial Hall (for the arts), the cricket club, and St Andrew's cathedral. These buildings surround what is now known as the *Padang*, a grassy area which started as a village green in the European residential section and later become a centre for colonial social and sporting life (B. Lee 1989).

The colonial regime reinforced ethnic residential and occupational segregation. The Europeans, displaced by the city residential pressures, moved northward and settled primarily in the Tanglin area, building spacious bungalows on elevated grounds. The establishment of the Tanglin Barracks and the relocation of the Botanical Gardens further reinforced the European character of Tanglin (Humphrey 1985). The Chinese and Indian traders, merchants, small shopkeepers, and labourers remained in the city. A large Chinatown developed adjacent to the mouth of the Singapore river and immediately north of the mercantile area. The Indians concentrated in the dock area, south-west of the Singapore river, and in pockets along High Street, Arab Street, and Serangoon Road (Hodder 1953). The Malays, who traditionally lived by fishing, were marginalized by their inability to compete in commerce with the immigrant groups. They generally stayed away from the city (Turnbull 1977), except for a pocket to the south-east in an area called Kampong Java adjacent to the Malay sultan's property. Here the Sultan mosque remains a powerful focus for the Malay community even today.

The failure of the colonial government to address social needs, particularly in the area of housing, created a dualistic structure in central Singapore. In place was the excellent entrepôt infrastructure (harbour, warehouses, storage, transport, and communication facilities), and various supporting services (banks, trading houses, wholesalers, and transport and communication services) that formed the backbone of the Singapore economy. This economic machinery was maintained within a deteriorating social landscape as rapid population growth led to high residential densities in the city's central area.

Contemporary Forces Shaping the Built Environment

The developmental state

In the 1950s and 1960s Singapore was in great political turmoil. Although the British had laid a constitutional framework and made some arrangements for transfer of power (Turnbull 1969), the struggle for power among local political parties created an instability that was reinforced by regional problems arising from Indonesia's domestic tensions, expansionist tendencies, and political differences with Malaysia.

In the early 1960s, the ruling political party, the People's Action Party (PAP), launched a series of initiatives and changes. Measures to increase government control were directed at the media, left-wing student groups, and trade unions—the prime sources of political challenge and civil disorder in that period (Gamer 1972). The state bureaucracy that had developed under the colonial regime was also transformed. New initiatives included a centre to reorient civil servants, a scheme to rationalize work practices, attempts to sideline more left-wing elements, and a working arrangement with moderates, who increasingly saw their interests as being aligned with the developmentalist programmes launched by PAP (Seah 1976). Most importantly, PAP left housekeeping functions to the civil bureaucracy and developed a new group of para-government statutory boards to manage specific developmental tasks. These included attracting foreign investments (Economic Development Board or EDB), improving public utilities (Public

Utilities Board or PUB), managing the port and coastal traffic (Port of Singapore Authority or PSA), renewing the urban environment (Urban Redevelopment Authority or URA), building public housing (Housing Development Board or HDB), and developing industrial estates (Jurong Town Corporation or JTC).

As a new political party without a track record, PAP needed to gain the confidence of the electorate. The party achieved this objective largely through the provision of public housing. The relationship between the success of the public housing scheme and PAP's popularity can be noted from the following account: 'Public housing is probably the most visible and demonstrative project in the Republic. Its success had assured support for many other government policies. It is perhaps not a pure coincidence that results of the study on the level of satisfaction with public housing are similar to the result of the 1972 election in which the People's Action Party received 70% of the popular vote and won all the seats in the parliament' (Yeh and Pang 1973, quoted by Hassan 1976a: 240). The success of public housing, which was made available for purchase, gave rise to a nation of home owners, creating in the minds and hearts of Singaporeans an 'increased ideological and material commitment to the system' (Chua 1991: 34). The public housing programme was also supplemented by a strong education and public health initiative, which improved the poor conditions that had existed under the colonial regime and also helped provide PAP with the legitimacy needed to launch other developmental programmes.

The expanding middle class and civil society

The great transformation in the first two decades of independence was largely a state-driven initiative. The economic rationale of such a transformation—better jobs, a higher standard of living, and higher wages—was essentially accepted by the people (Chua 1989). Being comfortable with the pace and the results of such a transformation was, however, another matter.

Three decades of development and growing affluence have created an expanding middle class of better-educated and vocal Singaporeans. It is this group which in the 1980s has started reassessing the state–society relations along a number of fronts. Appalled by the rapid and massive urban redevelopment schemes, a group of architects in the late 1970s and early 1980s worked through Singapore Institute of Architects, Singapore Institute of Planners, and Singapore Planning and Urban Research Group to mobilize support for a more sustained urban conservation effort (Wei 1993). The movement to conserve the built heritage of Singapore was conceived of in terms of the need to establish links with the past. The appeal to the middle class for support is clear from this comment made by a leading proponent of the movement: 'the young middle class already have their homes. The next best thing to fight for is their roots, some place in time and space that they can identify with' (Wei 1993: 20).

The 1980s also saw a better-organized and a more vocal women's movement. In 1980, the Singapore Council of Women's Organization was formed as a non-governmental umbrella organization co-ordinating the efforts of various local women's organizations. In 1985, the Association of Women for Action and Research (AWARE) was launched to promote greater awareness, increased participation, and equal opportunities for women (Wee 1987). AWARE, in particular, has been active in engaging the government on issues of medical benefits for women, child-care centres, and on legislating protection for rape and wife-battery victims.

Perhaps the most important event affecting the relationship between the middle class and the state was a dramatic 13 per cent loss suffered by the ruling party in the 1984 election. While it was not clear whether the middle class was largely responsible for the shift in voting patterns, the effect was a concerted effort by the government to consult the middle class in the policy process (Rodan 1996). A gradual shifting of the state position to consider middle-class views cumulated in two ministerial comments made in June 1991:

Once you reach a certain level of industrial progress, you've got an educated work force, an urban population, you have managers and engineers. Then you must have participation because these are educated, rational people. If you carry on with an authoritarian system, you will run into all kinds of logjams. You must devise some representative system . . . Then you may get the beginnings of a civic society, with people forming their own groups . . . Almost spontaneously, these will form, because being educated, knowing of the wider world, will bring like-minded people together. Then only do you have the beginnings of what I would call an active grassroots democracy. (Lee Kuan Yew, in an interview with *The Economist*, quoted in Jones and Brown 1994: 81)

Without a strong civic society, the Singapore soul will be incomplete. If the creation of a strong state was a major task of the last lap, the creation of a strong civic society must be a major task of the next lap . . . For our civic institutions to grow, the state must withdraw a little and provide more space for local initiative . . . The problem now is that under a banyan tree very little else can grow. When state institutions are too pervasive, civic institutions cannot thrive. Therefore it is necessary to prune the banyan tree so that other plants can also grow. (National University of Singapore Society Inaugural Lecture, 20 June 1991 (Yeo 1991: 80, 84))

While the terms 'civic' and 'civil' society may be open to alternative readings (Nair 1993), there is clearly an attempt by the state to allow a space for non-governmental organizations to participate in the policy process. In the 1990s, middle-class participation in the public forum has become a significant force shaping the built environment. Specifically, it is the middle class which has pushed civic and professional organizations such as the Singapore

Institute of Architects and the Nature Society (Singapore) to engage the government on its environment policies.

The changing economy and international relations

The regional instabilities of the 1960s no doubt influenced the government's decision to attract multinational corporations to Singapore rather than relying on a regional development strategy with Malaysia and Indonesia. By the end of the 1960s, Singapore was becoming increasingly attractive as a place for direct foreign investment. The local political situation had stabilized considerably with PAP's control of opposition groups and the party's increasing popular support as a result of its social programmes. Regional political uncertainties had eased with the formation of the Association of South-East Asian Nations (ASEAN), a forum that provided a healthy dialogue in the region.

The 1970s saw a rapid growth of direct foreign investment due to the initiatives launched by Singapore's newly formed statutory boards, which built upon the excellent commercial infrastructure left by the British. In 1975, for example, investment in Singapore accounted for 49 per cent of all direct foreign investment in Asia (Mirza 1986). Singapore's successful industrialization strategy was based on development as an export platform for manufacturing by multinational corporations. This approach strengthened Singapore's links to the global economy, and particularly to the United States.

The economic development of Singapore in the 1980s increased international and regional links. Faced with rising costs and shrinking land and labour resources, the government's economic strategy turned increasingly toward attracting and consolidating core economic activities (Arrighi and Drangel 1986). Two policies emerged in the mid- to late 1980s: establishing Singapore as the operational headquarters scheme for international corporations (Ho 1991; Dicken and Kirkpatrick 1991) and promoting regional development in partnership with Indonesia and Malaysia (Lee 1991).

The thrust of the state's economic initiative in the 1990s is to expand on Singapore's regional role through investments in overseas ventures in close alliance with local capitalists. If this plan succeeds, Singapore will become a command centre as production activities in the region become increasingly directed from Singapore (Friedmann 1986; Sassen 1991).

The increased international and regional linkages mean that Singapore's image abroad becomes even more crucial as a means of maintaining her economic position. Although colonial Singapore functioned as an international port city, its relations with the external world were managed by Britain. Now, as an independent city-state, Singapore makes its own foreign policy. And as the city's ties to the rest of the world expand, Singapore is expected to play a role as a good global citizen by advocating free trade, enforcing the international copyright law, maintaining a good human rights record, treating its workers well, and fostering environmental protection.

The built environment of post-colonial Singapore

By the end of the 1970s, Singapore's development programmes were yielding handsome dividends. The multinationals were bringing in capital, and industrialization was creating high rates of growth and full employment. The education system provided compulsory schooling and geared its students towards the industrial economy. Health conditions had improved drastically with a comprehensive system of government out-patient clinics and hospitals, and massive urban renewal had created a modern city with high-rise office buildings, shopping centres, and hotels.

The result of planning efforts was radically to reshape the city and the suburb. The proportion of the population living within the city declined from 63 per cent in 1957 to 60 per cent in 1970 and 52 per cent in 1980 (Humphrey 1985). As the entrepôt economy declined, import and wholesale activities were relocated outside the city, along with cottage and backyard industries, such as furniture-making, motor repair, and metalworking, which were moved to industrial estates (Cheng 1980). In place of these came the new economic foundations of the city. The 'golden shoe', Singapore's financial district, took shape in the late 1960s and the early 1970s, expanding from the heart of the traditional mercantile area at Raffles Place, along Shenton Way, Robinson Road, and Anson Road (URA 1989).

Retail activity also went through a dramatic transformation. By the 1970s, the more exclusive retail outlets moved into the shopping centres that clustered along Orchard Road and, on a smaller scale, into the suburbs (Sim 1984). Orchard Road itself underwent a major transformation in the 1970s, changing from an arterial road fronted by small retail and service shops to the 'main street of the nation' (Ho 1989) or the 'new city' (Ong 1982) packed with shopping centres interspersed with major hotels.

These changes to the city centre could only have occurred because of the parallel development that took place in the suburbs. Public housing and industrial estates absorbed the residential population and small businesses that were displaced by urban renewal programmes. Retailers specializing in the daily necessities, such as grocery shops, followed the population out of the city into the Housing Development Board (HDB) facilities. The industrialization programme's requirement for space and labour were being met by building industrial estates within and close to new towns.

For an increasing segment of the population, the 1970s marked a new era where a way of life was dramatically transformed by the move into public housing. Families in public housing estates were generally satisfied with improved amenities such as better access to schools, markets, and recreation. The disadvantages were higher housing costs compared with the older villages and the loss of a sense of solidarity, mutual help, and a warm social environment (Chang 1975; Hassan 1976*b*; Tai 1988). With a public housing policy of ethnic residential integration, HDB neighbourhoods also replaced the ethnic enclaves of colonial Singapore.

The transition from the old neighbourhoods to public housing estates was largely completed by the end of the 1970s. In 1980 close to 70 per cent of the population was living in public housing. A new generation has now grown up who consider HDB estates as their home. This sentiment, however, should not be taken to mean a commitment to place. Residential mobility is high: young people get married and move away from their parents, growing families move to larger units, and the wealthy choose private housing or newer estates with better amenities.

Singapore has experienced three decades of rapid and massive state-led development. The resultant economic growth created a middle class which has started reassessing the price of progress as affluence has weakened, in the minds of Singaporeans, the appeal of the development imperative. As an international city, Singapore is also increasingly open to the opinions of its trading partners. Within these changing circumstances, the debate between development versus nature conservation has assumed public importance.

Development versus Nature Conservation

Land use in Singapore is not generally politicized or open to public debate. Since 1967, the state has had legal power to acquire land and has exercised this right extensively for the last two decades, both for urban renewal projects in the city centre and for the construction of new industrial and housing estates in the suburbs. There are no institutional mechanisms, such as town meetings or community newspapers, that would allow for community interests to be focused and expressed. A member of parliament could potentially express the concerns of his or her constituency, but Singapore 'meet the people' sessions with parliamentarians tend to be dominated by individuals and are not conducive to expression for the collective voice.

One of the few instances of community mobilization resulted in the 1981 by-election of an opposition member to parliament, the first in more than ten years. Understandably, the PAP-dominated government is concerned about community relations. At the same time, social differentiation in the population along income, education, and ethnicity lines has reached a level of complexity that general policies that worked well in the past are no longer viable (Chua 1993).

To illustrate the political process and explain the mediation between economics and culture in the justification of land use, this chapter will focus on nature conservation in Singapore. More specifically, the issue of nature conservation addresses the following concerns in managing Singapore's built environment:

1. The tension between development versus conservation. Conservation in the 1990s involves protecting what is left of Singapore's natural environment from the state's development efforts. With land extremely scarce, the tension is high.
2. The role of fledgeling civic groups that are voicing concerns, gathering public support, and offering suggestions to the government on nature conservation.
3. The global–regional–local interface. Nature conservation has become an international issue, and both corporations and national governments alike feel pressure to adopt a pro-environment position to maintain an image of good global citizens.
4. A changing state–society relation on the issue of the environment, with public support building increasingly towards a pro-environment stance.
5. The evolution of a government policy incorporating nature conservation within a developmental-managerial approach.

Singapore's record on environment issues has been perceived positively (see Ooi 1992) as well as more critically (see Wee 1993). These mixed views may reflect the government's changing stance on environmental concerns, motivated largely by a desire to present a positive image to Singaporeans and to the rest of the world.

The PAP initiated action on both the environmental and developmental fronts in the 1960s when it assumed control of the government. The party's record on development is impressive; its success in managing the environment has resulted in Singapore being known as a 'clean and green' city. Although these two words have appeared alongside each other from the beginning, they stemmed from different concerns when they were first introduced. Of the two, cleanliness was more important in the 1960s. The emphasis on cleanliness was derived from the new government's preoccupation with eliminating indiscriminate littering in public places. A number of visitors to the colony had described the Singapore streetscape as smelly, dirty, unhygienic, noisy, overcrowded, and disorderly (see Savage 1992 and Tarling 1992 for examples). This situation was attributed to a failure by the colonial administration to keep public places clean. The emphasis on cleanliness was traced to a perception that dirt and disorder were tied to a Third World condition which must be eradicated: cleanliness was seen as a sign of a well-managed and ordered society. Such ideas are apparent in speeches made by Prime Minister Lee Kuan Yew in the 1960s (compiled in Koh 1976: 203, 206–7):

The problem is: how do you maintain, in a society of relatively low living standards packed in 224 square miles of island, standards of hygiene and public cleanliness equal to those of the West? We cannot carry on the old system. This was left to overseers and 'mandores' in a leisurely old colonial society. (Speech at an inaugural dinner organized by the University of Singapore Business Administration Society, 27 Aug. 1966)

We are one big urban area, one big city. If you slacken on discipline and allow standards to go down, then you have lost everything. You will make no progress. So despite our pressure of population, despite the fact that we must seek popular support every five years, we must enforce high standards ... Where do we start? First a good cleansing service and enforcement of minimum standards of social behaviour. (Closing ceremony of the People's Association Second Management Committee Conference, 21 July 1968)

'Clean' was thus seen as an essential condition marking Singapore's transition from the colonial era, and signifying to Singaporeans and outsiders alike a new management and order characterized by modernization and progress. Seen within this context, 'green' is a secondary but nevertheless important concern tagged on to 'clean', acting as an appendage, a dressing for the new order. The notion of 'green' emerged from a vision of Singapore as South-East Asia's garden city. Yeh (1989) documented how the idea was formalized, with an interministry Garden City Committee formed in the mid-1960s, a division overseeing parks and trees in 1968, and a national tree planting day in 1971.

It is important to note that, in the 1960s and 1970s, 'green' represented the government's commitment to creating a garden or park-like atmosphere within the city-state. It did not mean a commitment to conserve natural areas. The government's primary concerns during this period—to create industrial estates for foreign capital, a modern city centre for commerce and services, and a public housing programme—required a massive land-acquisition and development effort. These priorities required encroachment on wooded areas, swamps, coastal areas, and farmland. The rationalization for this exercise can be deduced from another speech made by Prime Minister Lee Kuan Yew:

With the rapidly expanding population, growing pace of industrialisation, and increasing tempo of housing and other public development, land is becoming a scarce and very expensive commodity in Singapore. It is imperative, therefore, that all available land is put to most advantageous use for the benefit of the people of Singapore. Swamps and waste lands, therefore, have to be reclaimed. So too suitable stretches of foreshore along the coast of Singapore. (moving the second reading of the Foreshores (Amendment) Bill, Legislative Assembly, 10 June 1964 (Koh 1976: 199))

This view was incorporated into the government's approach to development. In discussing land-use planning, the acting chief planner stated in the 1970s that in Singapore 'the overriding importance of industrial development is unquestioned' and reiterated the need to reclaim inland and coastal swamps for industrial use (Chua 1976: 174, 176). The developmentalist orientation of the government meant that land was valued as a factor of production. Undeveloped land was 'waste' land and its utility was only realized when rent could be extracted. In this context, the essential idea behind the garden-city concept was to avoid the concrete-jungle effect of an urban environment by landscaping with trees and shrubs, rather than keeping the environment in its natural state. Kong and Yeoh (1992: 37) observed that 'while on the one hand, natural areas continued to be destroyed, on the other hand, various policies and actions have been introduced to green the city'. This is not really a paradox: economic concerns required the acquisition of all 'unproductive' land for redevelopment, while managerial concerns required that the transformed landscape conformed to standards of efficiency, cleanliness, and attractiveness. Given these standards, gardening the landscape made sense. The concern to create and maintain a clean and beautiful city was also supported by a number of anti-littering and anti-pollution laws.

This approach may have had positive effects in terms of the capacity of planted trees and shrubs to produce oxygen (the notion of 'green lungs' was used prominently in the government's campaign), provide shade, and reduce temperatures in tropical Singapore. Given the bureaucratic approach, however, the tendency was for the mass plantings of a few species that adapt well to urban conditions. The use of insecticides to keep the plants thriving had the unintended effect of disrupting the food chain, particularly for birds (Wee 1993). The absence of plants that produce fruits and berries also tended to reduce bird life. In addition, grass cutting and other regular forms of garden and park maintenance constrained the breeding of birds and other animals (*Straits Times*, 3 Nov. 1991).

By the 1970s, Singapore had reached a momentum in cleaning and gardening the city. In place were the institutions responsible for the task, a public awareness programme, and a set of financial disincentives to deter 'antisocial' behaviour. Public tastes and perceptions were also shaped by these state initiatives. As highlighted by the director of the National Parks Board (*Straits Times*, 21 Apr. 1991), because of a successful garden-city campaign, Singaporeans became used to manicured gardens and neat rows of trees, rather than natural environments such as forest, grasslands, and swamps. Thus, with the public socialized to prefer landscaped gardens, public opinion supported existing state practices in land use.

As a result of various development schemes, between 1967 and 1982 there was a 29 per cent reduction in swamps, fresh-water areas, and wooded areas (Ministry of National Development 1983). Of the five nature preserves established by the colonial government, Pandan Nature Reserve, protected in 1884, was reassigned for industrial purposes in 1962, Kranji Nature Reserve was redeveloped as a reservoir in 1973 (Wee 1993), and land reclamation and other development led to the reclassification of Labrador Reserve into a park. The two remaining reserves were also damaged by development. The Central Water Catchment Area suffered encroachment and environmental damage through reservoir construction, development of the Singapore Zoo, introduction of an arms-manufacturing facility, and other military uses. The Catchment Area was separated from the Bukit Timah Reserve in the 1980s with the construction of an expressway. This separation blocked the movement of animals between the areas, reducing variety in their gene pools. The expressway also affected drainage patterns in adjacent areas (Interministry Committee 1992). In late 1993, the Environment Minister was still reminding the public of the government's emphasis on the development–environment equation: 'While Singapore must not achieve economic progress at the expense of the environment, environmental considerations may at times have to be overruled where there are clear overriding benefits to the people and the country' (*Straits Times*, 10 Nov. 1993).

Nevertheless, between the 1970s and the 1990s, there was a gradual shift toward environmental concerns. The clean-up of the Singapore and Kallang rivers represented the beginning of such a shift. The river clean-up campaign spanned a period of ten years (1977–87), and resulted in the relocation of 26,000 families, 4,296 street hawkers, and 2,800 backyard trading establishments. Altogether, 21,000 sites that lacked sewerage facilities were phased out (*Straits Times*, 10 No. 1990). The primary motivation was not environmental protection, but, as the Prime Minister indicated in 1977, Singapore's need for potable water (Hon 1990). The impact of this massive project, however, provided several conditions for changing state and public attitudes about the environment. First, the Ministry of Environment (ENV) managed this project rather than the Urban Redevelopment Authority (URA). As a result, the success of the river clean-up was viewed as an environmental rather than a development achievement. This success provided a platform for ENV to expand its activities. Second, the project introduced the concept of conservation to a broad segment of the bureaucratic élite. In addition to the ENV, the project involved every statutory board with responsibilities for Singapore's built environment. This was because the resettlement of boat owners, farmers, small businesses, and residents required the co-operation of all these agencies. In the clean-up process, there were several competing interests, for example, between the Economic Development Board (EBD) and ENV over the removal of boatyards in the Kallang Basin (Hon 1990). The resolution of such conflicts enabled an embryonic consensus to emerge. The ten-year time-span for the project allowed the ENV to work out co-operative arrangements with these organizations.

Third, the remarkable transformation of the Singapore and Kallang rivers gave the state two monuments to showcase its achievements. These became foundations on which to build the image of environmental responsibility. Both featured prominently in *Singapore's National Report* (Interministry Committee 1992) and the *Green Plan* (Ministry of Environment 1992), two official publications documenting the government's commitment to the environment. These symbols of environmental protection join the nature reserves in signifying the government's achievements. Although these places are 'natural', they should be considered as part of the 'built' environment since these have been appropriated by the government, are protected from other encroachments, and play an important ideological role. The symbolic value of the two rivers is reinforced by their proximity to the city centre. As central places, these have a tendency to stay in people's memory. The Singapore river has the added primordial symbolic power of reflecting the economic origins of the city-state. Three years before the end of the clean-up, the Urban Redevelopment Authority (URA) had already appropriated this image in its attempt at representing Singapore city: 'Various cities of the world all have memorable spaces. Singapore is unique in that the biggest area of public space will be a body of water [referring to the Singapore river mouth area]. Singapore will be remembered by this body of water' (*Straits Times*, 16 Dec. 1984). And fourth, the project demonstrated to the public the value of conservation by bringing filthy rivers back to life. The government has used the success of the river clean-up for ideological purposes, for example, giving prominence to the rivers in television footage featuring Singapore's national day songs. This, in turn, reinforced the value of conservation.

The river clean-up project established a moral (the value of conservation in the eyes of the public) and ideological (the government's use of conservation as a new way of reinforcing its legitimacy in Singapore and the world) front. Soon afterwards, the government accepted a proposal from the Nature Society (Singapore) (NSS) to designate Sungei Buloh as a bird sanctuary. As Wee (1993) pointed out, this was the first major concession to nature conservation after a twenty-three-year history characterized not only by the destruction of natural land and marine habitats but also by encroachment on gazetted nature reserves in the interest of development. The government's acceptance of the proposal also lent legitimacy to the NSS, opening a role for non-government organizations to participate in decision-making on land-use policies.

The designation of Sungei Buloh as a bird sanctuary did not set a larger trend. The NSS made a similar call to preserve the 215-hectare Kranji marshes. This time, however, the request was turned down by the Ministry of National Development, which had allocated large tracts of the marshes to Singapore Telecom, Singapore Broadcasting Corporation, and the Public Utilities Board (*Straits Times*, 15 Oct. 1990).

The government proposal to build a golf course at the Lower Peirce Nature Reserve in 1992 saw the first sustained engagement between state and civil society on nature conservation. Most significantly, this episode represented one of the few times when the government actually backed off from a proposed development because of public opinion. The outcome of this episode had, in fact, been influenced by events that occurred several years earlier.

In 1990, the United Nations honoured Singapore by appointing a Singaporean, Tommy Koh, to chair its Preparatory and Main Committees for the International Conference on the Environment and Development (popularly known as the Earth Summit) held in Rio de Janeiro in 1992. Professor Koh told reporters that 'Singapore was chosen because of the high environmental standard it was able to uphold while achieving 25 years of rapid economic progress' (*Straits Times*, 7 Mar. 1990). When the NSS produced the environmental impact assessment report on the proposed golf course development at Lower Peirce, Professor Koh added his weight by writing a foreword that summarized the project's potential damage to the environment (Nature Society (Singapore) 1992).

Given Singapore's role in the Earth Summit and nudged on by the momentum gained from earlier government initiatives, the government embarked in the 1990s on further pro-environment programmes. The government designated Singapore as an 'environmental city' for the region and formulated an international environment policy in *Singapore's National Report* (Interministry Committee 1992) and the *Green Plan* (Ministry of Environment 1992). In 1991 a conservation area was designated to protect mature trees and a reforestation programme was initiated for Singapore's nature reserves (*Straits Times*, 28 Dec. 1990 and 7 Aug. 1991; *Business Times*, 4 Nov. 1991). The creation of a new division for environmental planning and conservation in the Ministry of Environment (ENV) opened a wedge within the vast state bureaucracy and provided a check against pro-development interests. In the past, ENV's objectives of controlling pollution and maintaining a park-like environment were in harmony with the pro-development stance adopted by other governmental agencies. By the 1990s, ENV's increasingly pro-green position, although popular in regional and international circles, created tension with pro-development forces at home.

In 1990, journalists in the broadcasting and the print media formed the Singapore National Forum of Environmental Journalists (*Straits Times*, 19 Dec. 1990). Their goal was to direct greater media attention to environmental issues, and resulted in greater public awareness of environmental issues.

These measures played a crucial role in shaping public opinion on nature conservation and public perception on the government's pro-environment stance and decisively influenced the outcome of the 1992 government proposal to convert part of the Lower Peirce Nature Reserve to build a golf course.

The proposal was announced at a time when Singapore was enjoying a high profile at the Earth Summit. A public debate in the newspapers and magazines then followed. An early letter by Chen mentioned that the government's golf course proposal was ironic since 'the parties involved are the same people talking about saving the earth and protecting the forests', suggesting the apparent contradiction between Singapore's role in the Earth Summit and its actions at home. Chen advocated that Singapore's nature reserves must be made a 'sacred cow'.

The intensity of public opinion on this issue took everyone by surprise. In *Calibre*, a magazine for professionals, Betty Khoo described public response as vehement, and overwhelmingly against the government plan. She went on to describe the feelings of anger, frustration, and disgust expressed by the people she interviewed (Khoo 1992). The Singapore Institute of Landscape Architects joined the NSS in urging that the Lower Peirce Nature Reserve be preserved, arguing that 'we should not view nature reserves as a land bank reserved for future development, but rather as a permanent national asset'. It also stressed that public opinion should be invoked to determine whether the project should proceed or be aborted (*Straits Times*, 1 June 1992). This view represented an attempt to shift authority for an important land-use decision from the government to the people. In response, the government emphasized the importance of expert opinion and the danger of public sentiment: 'Proposals for nature conservation must be made in a proper perspective. In land-scarce Singapore, we must use land optimally . . . We would be guilty of poor judgement and the nation would lose if worthy projects were aborted because of misguided or irrational pressures' (*Straits Times*, 6 June 1992). This set of responses also shows the interplay between culture and politics. Letter-writer Chen and the Landscape Architects called for alternative cultural interpretations. The terms 'sacred cow' and 'permanent national asset' suggest not only an alternative view of nature reserves, but a view that they need to be elevated to the point of being valued as sacred and priceless.

Two months after the project first came to public attention, it was raised in parliament, and the Minister for National Development replied that the golf course was 'just an idea', and would not be approved if it would cause extensive damage (*Straits Times*, 1 Aug. 1992). Two days later, the Deputy Prime Minister told Singaporeans not to become emotional about protecting the environment. His reported comment was that 'the golf course would have to be built on nature reserve if there was a need' (*Straits Times*, 3 Aug. 1992).

The golf course project has been shelved for the time being. Educated by the media and influenced both by the government's pro-environment initiatives and by developments overseas, the public has moved to a pro-environment position. A recent public opinion poll indicated that Singaporeans have become increasingly aware of the need to protect the environment and want the government to give environmental issues a high priority (*Straits Times*, 19 Oct. 1992). The public is likely to object if the golf course project is revived.

The NSS, confident of increasing public support, will continue to be Singapore's watchdog on environmental conservation, and other organizations will undoubtedly join the cause. In this situation, the government will probably find it difficult to proceed with projects that jeopardize the environment. In addition to local public opinion, government initiatives in this area will be constrained by the desire to protect Singapore's pro-environment image overseas. At the bureaucratic level, conflicts are likely to emerge between agencies charged with projecting Singapore's environmental image and those that are concerned more narrowly with development.

Conclusion: State, Society, and the Built Environment

Singapore's colonial government, driven by commercial interests, developed a settlement around the harbour and expanded the port infrastructure throughout the nineteenth and early twentieth centuries. Suggestions for manufacturing were brushed aside in favour of commerce, and social expenditures were ignored, leaving ethnic communities to fend for themselves. With these policies, the colonial government created a dualistic structure, where a well-developed commercial infrastructure coexisted with deteriorating living conditions.

The post-independence government adopted new policies to strengthen social services in the areas of housing, health, and education, and to broaden Singapore's economic base by diversifying into manufacturing. The success of such programmes provided the popular support necessary for the state-controlled development efforts. Government policies led to the rapid transformation of Singapore's built environment, resulting in a city centre dedicated almost exclusively to service industries, combined with public housing and industrial estates in the suburbs.

The recent development–environment debate in Singapore illustrates how state control of the built environment may become increasingly open to public influence. The process shows the interplay between political, economic, and cultural forces. In the 1960s and 1970s, economic development was the main basis for government land-use decisions. Successful government programmes to develop and manage the environment also shaped public tastes and perceptions. As a result, gardens and parks were valued more than natural reserves such as forest, swamps, and grasslands.

The shift towards a greater concern for natural environments began with the clean-up of the Singapore and Kallang rivers. Together with the nature reserves, the rivers emerged in the 1990s as symbols of the government's environmental achievements. Singapore's role in the Earth Summit further strengthened the value of conservation at home as a means of enhancing the country's image. Inevitable conflicts between developmental and environmental goals came to the fore in the controversy over a proposed golf course development in the Peirce Nature Reserve. This episode demonstrated how different cultural values affect land use—developmentalist versus environment and gardening versus conservation. It also illustrated changes in the role of public opinion in Singapore and the shifts within different agencies of the government bureaucracy.

References

Allen, D. F. (1950). *Report on the Major Ports of Malaya.* Kuala Lumpur: Government Press.

Arrighi, Giovanni, and Drangel, Jessica (1986). 'The Stratification of the World-Economy: An Exploration of the Semiperipheral Zone', *Review*, 10: 9–74.

Bellet, J. (1969). 'Singapore's Central Area Retail Pattern in Transition', *Journal of Tropical Geography*, 28: 1–16.

Castells, Manuel (1981). 'Towards a Sociological Theory of Planning', in Charles Lemert (ed.), *French Sociology: Rapture and Renewal since 1968*. New York: Columbia University Press.

Chang, Chen-Tung (1975). 'A Sociological Study of Neighbourliness', in Stephen H. K. Yeh (ed.), *Public Housing in Singapore*. Singapore: Singapore University Press.

Cheng, Lim-Keat (1980). 'Changing Patterns of Spatial Organization in Singapore', *Journal of South Seas Society*, 35: 59–71.

Chua, Beng-Huat (1989). 'The Business of Living in Singapore', in Kernial Singh Sandhu and Paul Wheatley (eds.), *Management of Success*. Singapore: Institute of Southeast Asian Studies.

—— (1991). 'Not Depoliticized but Ideologically Successful: The Public Housing Programme in Singapore', *International Journal of Urban and Regional Research*, 15: 24–42.

—— (1993). 'Beyond Formal Strictures: Democratisation in Singapore', *Asian Studies Review*, 17: 99–106.

Chua, Peng-Chye (1976). 'Optimisation of Shrinking Land Resources in Singapore', in John Wong (ed.), *The Cities of Asia*. Singapore: Singapore University Press.

Dicken, Peter, and Kirkpatrick, Colin (1991). 'Services-Led Development in ASEAN: Transnational Regional Headquarters in Singapore', *Pacific Review*, 4: 174–84.

Dixon, Chris (1991). *Southeast Asia in the World Economy*. Cambridge: Cambridge University Press.

Duncan, James, and Ley, David (1993). 'Introduction: Representing the Place of Culture', in James Duncan and David Ley (eds.), *Place/Culture/Representation*. New York: Routledge.

Eckstein, Susan (1990). 'Poor People versus the State and Capital: Anatomy of a Successful Community Mobilisation for Housing in Mexico City', *International Journal of Urban and Regional Research*, 14 (2): 274–96.

Fraser, J. M. (comp.) (1949). *The Work of the Singapore Improvement Trust, 1927–1947*. Singapore: Singapore Improvement Trust.

—— (1955). 'The Character of Cities: Singapore, a Problem in Population', *Town and Country Planning*, 23: 505–9.

Friedmann, John (1986). 'The World City Hypothesis', *Development and Change*, 17: 69–83.

Gamer, Robert E. (1972). *The Politics of Urban Development in Singapore*. Ithaca, NY: Cornell University Press.

Harvey, David (1973). *Social Justice and the City*. Baltimore: Johns Hopkins University Press.

Hassan, Riaz (1976*a*). 'Public Housing', in Riaz Hassan (ed.), *Singapore: Society in Transition*. Kuala Lumpur: Oxford University Press.

—— (1976*b*). 'Symptoms and Syndrome of the Developmental Process', in Riaz Hassan (ed.), *Singapore: Society in Transition*. Kuala Lumpur: Oxford University Press.

Ho, Jason W. P. (1989). 'Main Street of the Nation', Honours thesis, Department of Sociology, National University of Singapore.

Ho, K. C. (1991). *Studying the City in the New International Division of Labor*. Department of Sociology Working Papers 117. National University of Singapore.

—— (1993*a*). 'Industrial Restructuring and the Dynamics of City-State Adjustments', *Environment and Planning A*, 25: 47–62.

—— (1993*b*). 'Issues on Industrial and Urban Development in Local Literature: Public Housing in Singapore', in Boon-Hiok Lee and K. S. Susan Oorjitham (eds.), *Malaysia and Singapore: Experiences in Industrialization and Urban Development*. Kuala Lumpur: University of Malaysia.

Hodder, B. W. (1953). 'Racial Groupings in Singapore', *Journal of Tropical Geography*, 1: 25–36.

Hon, Joan (1990). *Tidal Fortunes*. Singapore: Landmark Books.

Humphrey, John W. (1985). *Geographic Analysis of Singapore's Population*. Census monograph 5. Singapore: Department of Statistics.

Interministry Committee (1992). *Singapore's National Report for the 1992 U. N. Conference on Environment and Development Preparatory Committee*. Singapore: Interministry Committee for the UNCED Preparatory Committee.

Jones, D. M., and Brown, D. (1994). 'Singapore and the Myth of the Liberalizing Middle Class', *Pacific Review*, 7: 79–87.

Khoo, Betty (1992). 'Enough Golf Courses. Save our Forest!, *Calibre*, July: 72–5.

Koh, Douglas (comp.) (1976). *Excerpts of Speeches by Lee Kuan Yew on Singapore 1959–1973*. Singapore-Malaya Collection: National University of Singapore.

Kong, Lily, and Yeoh, Brenda (1992). 'The Practical Uses of Nature in Urban Singapore', *Commentary*, 10: 36–44.

Lawrence, D. L., and Low, S. M. (1990). 'The Built Environment and Spatial Form', *Annual Review of Anthropology*, 19: 453–505.

Lee, Bonita M. S. (1989). 'Alien Culture, Native Land.' Honours thesis, Department of Sociology, National University of Singapore.

Lee, Edwin (1989). 'The Colonial Legacy', in Kernial Singh Sandhu and Paul Wheatley (eds.), *Management of Success*. Singapore: Institute of Southeast Asian Studies.

Lee, Tsao-Yuan (ed.) (1991). *Growth Triangle: The Johor–Singapore–Riau Experience*. Singapore: Institute of Southeast Asian Studies and Institute of Policy Studies.

Ministry of Environment (ENV) (1992). *The Singapore Green Plan*. Singapore.

Ministry of National Development (1983). *1982 Land and Building Use: Report of Survey*. Singapore.

Mirza, Hafiz (1986). *Multinationals and the Growth of the Singapore Economy*. London: Croom Helm.

Nair, Sheila (1993). 'Political Society', *Commentary*, 11: 15–19.

Nature Society (Singapore) (1992). *Proposed Golf Course at Lower Peirce Reservoir: An Environmental Impact Assessment*. Singapore.

O'Connor, Richard A. (1983). *A Theory of Indigenous Southeast Asian Urbanism*. ISEAS Research Notes and Discussions Paper 38. Singapore: Institute of Southeast Asian Studies.

Ong, Teng-Cheong (1982). 'Public Transportation: Where do We Go from Here?', in S. Jayakumar (ed.), *Our Heritage and beyond*. Singapore: National Trades Union Congress.

Ooi, Giok-Leng (1992). 'Harmonizing Environment and Development: Ideological and Other Considerations', *Southeast Asian Journal of Social Science*, 20 (1): 107–12.

Peterson, Paul E. (1981). *City Limits*. Chicago: University of Chicago Press.

Redfield, Robert, and Singer, Milton (1954). 'The Cultural Role of Cities', *Economic Development and Cultural Change*, 3: 53–73.

Rodan, Gary (1996). 'Class Transformations and Political Tensions in Singapore's Development', in Richard Robison and David S. G. Goodman (eds.), *The New Rich in Asia*. London: Routledge.

Sassen, Saskia (1991). *The Global City*. Princeton: Princeton University Press.

Savage, Victor (1992). 'Street Culture in Colonial Singapore', in Beng Huat Chua and Norman Edwards (eds.), *Public Space: Design, Use and Management*. Singapore: Centre for Advanced Studies and Singapore University Press.

Seah, Chee-Meow (1976). 'The Singapore Bureaucracy and the Issues of Transition', in Riaz Hassan (ed.), *Singapore: Society in Transition*. Kuala Lumpur: Oxford University Press.

Sim, Loo-Lee (1984). *Planned Shopping Centres in Singapore*. Singapore: Singapore University Press.

Stockwell, A. J. (1986). 'British Imperial Strategy and Decolonisation in Southeast Asia 1947–1957', in D. K. Bassett and V. T. King (eds.), *Britain and Southeast Asia*. Centre for Southeast Asian Studies Occasional Paper 13. Kingston upon Hull: University of Hull.

Tai, Ching-Ling (1988). *Housing Policy and High-Rise Living: A Study of Singapore's Public Housing*. Singapore: Chopmen.

Tarling, Nicholas (comp.) (1992). *Singapore and the Singaporeans since 1819*. Centre for Asian Studies Resource Papers 3. Auckland: University of Auckland.

Turnbull, C. Mary (1969). 'Constitutional Development 1819–1968', in Jin-Bee Ooi and Hai-Ding Chiang (eds.), *Modern Singapore*. Singapore: Singapore University Press.

—— (1977). *A History of Singapore: 1819–1975*. Kuala Lumpur: Oxford University Press.

URA (Urban Renewal Authority) (1989). *The Golden Shoe: Building Singapore's Financial District*. Singapore.

Webster, Anthony (1987). 'British Export Interests in Bengal and Imperial Expansion into Southeast Asia, 1780 to 1824: The Origins of the Straits Settlements', in Barbara Ingham and Colin Simmons (eds.), *Development Studies and Colonial Policy*. London: Frank Cass.

Wee, Vivienne (1987). 'The Ups and Downs of Women's Status in Singapore: A Chronology of Some Landmark Events (1950–1987)', *Commentary*, 7: 5–12.

Wee, Yeow-Chin (1993). 'Coping with nature and Nature Conservation in Singapore', in Clive Briffet and Sim Loo-Lee (eds.), *Environmental Issues in Development and Conservation*. Singapore: Singapore National Printers.

Wei, M. H. (1993). 'Urban Conservation: Beauty or Beast?' Honours thesis, Department of Sociology, National University of Singapore.

Wong, Lin Ken (1960). 'The Trade of Singapore 1819–1869', *Journal of the Malayan Branch of the Royal Asiatic Society*, 33 (192, part 4): 1–315.

Yeh, H. K. Stephen (1989). 'The Idea of the Garden City', in Kernial Singh Sandhu and Paul Wheatley (eds.), *Management of Success*. Singapore: Institute of Southeast Asian Studies.

—— and Pang, E. F. (1973). 'Housing, Employment and National Development: The Singapore Experience', *Asia*, 31: 8–31.

Yeo, Y. B. George (1991). 'Civic Society: Between the Family and the State', *Speeches* (*A Bimonthly Selection of Ministerial Speeches*), 15 (3): 78–86.

12
Culture and Urban Future in East Asia

Mike Douglass and Won Bae Kim

Introduction

In focusing on the interrelationship between culture and the city, we have emphasized a broad perspective that looks beyond short-term urban dynamics. History has been brought in to provide a deeper understanding for urban changes occurring in East Asia. Whereas much of recent literature on the East Asian cities has stressed population and economic concentration and resultant policy issues, the theme of urban transformation in cultural context has been taken up in this book, with the case study chapters revealing the interplay of culture, politics, and economics in creating and allocating space in cities. The whole structure of the book has been based upon a premiss that solutions to mounting urban problems are incomplete and ineffective without an understanding of historical and cultural dynamics beneath the surface of the contemporary city.

Analysing the historical and cultural dynamics at the city level, we have adopted an approach that explains the urban process as one of creating and transforming urban space (the built environment) and of using this space to sanction the distribution of social power as well as to provide the functional support for production, circulation, and accumulation of material wealth. This process is viewed as occurring through three mutually mediating dimensions: culture, power, and the built environment. Social power, defined as the relative capacity to exercise control over social relations and resources, thus conditions and is conditioned by rules of behaviour defined by cultural norms and institutions, just as it also enhances and is enhanced or constrained by the functional capacity and symbolic content of the built environment.

In this matrix of mutual interaction are the ideologies supporting the legitimacy accruing to power holders and power seekers; the concrete forms of social and political practices; the continuity of institutions through time that rests on collective memories and shared meanings partly drafted into the built environment; the physical and technical capacity to extend allocative and coercive power; and the social capacity to envision and collectively respond to desires and pressures for change. The physical outcome—the city—once formed, also forms human events, just as these events reshape the urban landscape (Pred 1986; King 1989).

In East Asia, the state has been the ultimate arbiter, and its role in society has been pervasive. Recent debates about the role of the state in achieving economic miracles in East Asia testify to its power and position. Equally importantly, the state, as a provider of institutional structure and allocator of resources, has dominated the urban scene in East Asia from feudal to modern times. The case studies on Beijing, Seoul, and Tokyo demonstrate the pervasive role of the state in creating and transforming urban space, in embedding and manipulating symbolism in urban space, whether it is Tiananmen Square, Sae'oon Sang'ga (a commercial complex), or the Imperial Palace. Even the cities of a more recent vintage such as Hong Kong and Singapore represent the ideology or vision espoused by the state. Public housing schemes in Hong Kong and Singapore are within the purview of the state and represent both the symbolic element of the developmental state and the functional purposes of reproducing labour power and diffusing potential social instability by housing workers and emerging middle classes for export-led industrial development.

However, significant social changes have been occurring as East Asian societies enter into a more prosperous stage of development. The rise of business and commercial interest has been phenomenal. The state, because of the very emphasis on economic development, is wrapped up with these élites and finds it difficult to disentangle itself from the nexus of state–economy interdependencies. The position of the state is also challenged by its very economic success: the rise of middle classes. Urban middle classes from Seoul down to Singapore have begun to exercise their power and make their voices heard on issues ranging from park provision to housing development. The creation and allocation of urban space is no longer solely the purview of the state but is contested by diverse interests, whether they be civil associations, interest groups, or individuals.

This contest over urban space is not limited to the more prosperous cities of East Asia but also extends to the cities in transition such as Beijing and Hanoi. The dilemma for Beijing and Hanoi is how to reconcile two different principles of socialism and market in the use and transformation of urban space. Socialism, and therefore the role of state, has to be redefined in so-called socialist market economies. Establishing an appropriate balance between moral and material incentives is proving to be extremely difficult, if not impossible. Unreined passion for profit is feared in Hanoi, especially with regard to the preservation of the Thirty-Six Ancient Streets quarter. Even though the power of the state is dominant in the streets of Beijing, reaction to corruption and increasing social inequality creeps into the centre of the city.

The City in Global–Local Interaction

Given these observations on recent changes in the urban process, the final topic to be considered is culture and the urban future in East Asia. As we set out our approach in the Introduction (Ch. 1 above), one of the major themes of this book was the city in a new global order. To reiterate, we took a view that urbanization and its human consequences are increasingly contingent upon the interaction between each society and the global capitalist system. We also emphasized the importance of culture in the dynamics of urbanization and the creation of the built environment of cities, and the role of cities in the localization of external influences. Going beyond both the deterministic world systems perspectives and overly generalized Asian models of development, the chapters in the book have attempted to localize the understanding of trends and impacts of changes taking place in an increasingly globalized economy.

A global–local framework allows us to look into the future of cities and the role of culture in urban transformation in East Asia. This framework also suggests that while recognizing some commonalities between cities in their response to global pressures, there is much more room for alternative paths of development than received structuralist or modernist paradigms have allowed. The purpose of the discussion here is not to draw pictures of the future of cities. Rather, it is to explore the ways in which cultural dimensions of these issues have emerged in East Asia at this critical historical juncture at which the post-Second World War epoch of halcyonic economic growth and authoritarian states appears to have run its course, and new ways are being sought to resolve both internal and external sociopolitical and economic dilemmas.

As noted above, fundamental shifts within and among the elements of the urban process over the past decade have changed both the urban and national development trajectories in East Asia. On the international scale, economic and political power is increasingly concentrated in transnational enterprises and networks, which directly through asset ownership and indirectly through licensing, subcontracting, and control of distribution and information channels now account for more than half of all the international flow of commodities and services.

As transnational capital moves through circuits of trade, finance, and production across national territories, its requirements for new types of urban spaces have been as profound as the economic restructuring that it has induced (Scott and Storper 1992). Among these requirements have been the creation of world finance and commercial centres, mega-infrastructure such as deep-draught harbours for a new generation of container ships and world-class airports for air cargo services, and, in newly industrializing countries such as China and Vietnam, the dedication of large urban sites to export-processing zones for foreign investment. New

towns and massive housing development have paralleled these developments. Since the mid-1980s, the transnationalization process has begun
to occur from within all of the more advanced economies of East Asia as
well, further reducing the leverage of governments, labour, and citizens
alike over economic enterprises.

At the national scale, in all but a few societies the most pronounced
change in relations of power over the past decade has been the success of
political reform movements. The establishment of civilian governments in
Korea and Taiwan, the eroded hegemony of the ruling party in Japan, and
the imminent end of colonial rule in Hong Kong as China itself continues to
decentralize political power are fundamentally diminishing the high degree
of state autonomy that was linked to economic growth in the recent past
(Douglass 1994). These movements have been closely connected to the
emergence of middle classes, which has been prompted by rapid industrialization and economic development. Each in its own way, East Asian
societies are engaged in a search for new economic and political bases to
bring them into the next century.

The acting out of these major stresses in East Asia will substantially fall
on the local, urban scale where, with the exception of China, three-quarters
or more of national populations now reside. Cities will not only have to
cope with the usual stress on infrastructure and services, but will also
confront rapidly shifting economic fortunes as investments switch across
regional, national, and international space with ever-increasing velocity.
International and transnational migration will also inevitably bring multicultural dimensions and cultural diversity to societies now imaging
themselves as racially and culturally homogeneous.

From a cultural perspective, four prominent dimensions cut across the
issues and dilemmas of the future of cities in East Asia. They are:

- increasing multiculturalism and transnational cultural networks;
- the commodification of culture and its representations in the built
  environment as a new source of comparative advantage in economic
  growth and industrial location;
- culture as both resilient and resistant force to global penetration; and
- culture as the basis for collective action toward social, political, and
  economic sustainability of urban development.

Multiculturalism and Transnational Cultural Networks

One of the most difficult issues that societies in East Asia face is the
realization that international migration is not only increasing in scale and
range, but will be a permanent feature of their major cities in the twenty-
first century. While many governments adopt policies that try to wish this

reality away through immigration laws that seek to prevent rather than accommodate it, the ever-widening income gaps between richer and poorer countries and ease of international migration result in statistics showing steadily increasing numbers of legal and illegal migrants in East Asian cities even in situations of economic slow-down, such as in Japan (SOPEMI 1993; Pang 1993; Kim 1995).

The successes of the East Asian economies are accompanied by increasing international migration to them, and recent data now show that the industrialized countries of East Asia are absorbing increasing shares of international migration, albeit still on a relatively modest level. As with situations in the rest of the world, illegal migrants face the worst forms of exploitation, social discrimination, and, almost by definition, harassment by governments. They are without basic legal rights or protection in most instances, even in cases of severe injustices that contravene the laws of the host country. Yet despite these difficulties, migrants still attempt to settle into cities, to form 'ethnic' neighbourhoods, and to carry on with the raising and education of their children (Machimura 1993; Javier 1993).

How East Asian societies will deal with these growing numbers of people will be a critical element of their urban habitats and economies in the coming century. Whether it is maids in Hong Kong and Singapore, women brought for prostitution into Japan, or low-wage blue-collar workers in Taiwan and Korea, if host societies continue to ignore the reality of this global process, the result may well be the formation of foreign ghettos that will, in some instances, reach very large proportions of major cities. At a very general level, the outcome may not be too dissimilar to that of large cities in the West where the inequalities associated with multi-ethnicity and social distances among groups are among the most socially volatile issues of these times. If, on the other hand, they choose to accommodate and open cities to more inclusive, less discriminatory access to housing, land, urban services, work, and basic legal protection, a more socially equitable and culturally rich process of urban growth may be achieved.

In the contemporary era of advanced global transportation and communications technologies, international migration from lower- to higher-income countries and cities is experiencing a new cultural dimension, namely the appearance of what some have called 'transnational migrants' (Schiller *et al.* 1992). These migrants are part of a new type of culture–city relationship that is increasingly becoming the hallmark of the Pacific Rim. Their situations cannot be captured by conventional categories of permanent migrants, return migrants, or sojourners. They are, for the most part, members of two or more societies and often move freely among them. They thus represent more than just migrant labour; they sustain ties to their home societies and their families tend to remain functional across national boundaries (Jones 1992).

In this regard, the Korean community in the United States, which has simultaneously brought Korean cultural influences to such cities as Los Angeles and has undoubtedly become part of a network transmitting new cultural practices back to Korea, is striking—if for no other reason than that Seoul and Los Angeles arise from such different cultural roots. Los Angeles, the world is reminded, was nurtured as a grand theme park, a place where history can be instantly fabricated and where, in the post-modern sense of the urban condition, the image seems to be more important than the reality (Soja 1989). Yet it is this city where several hundred thousand Korean people have transplanted themselves to form 'Korea Town' and have become part of suburban America as well. Seoul, as much as Los Angeles, is part of this transnational culture. How it will deal with these linkages to cultural shifts signalled by its diaspora may well prove to be one of the more decisive questions in the future of cities throughout the world, including those in East Asia.

In the past, the Korean government has attempted directly to suppress and control these linkages on both ends of the relationship by trying, first, to limit travel outside of Korea and, secondly, to prevent US Korean-language newspapers and other sources of information from 'dissidents' abroad from appearing in Korea. Like many other efforts, these too seem to have been discredited under the new spirit of openness and democratic governance. If so, the 'Orange Generation' phenomenon—supposedly Western (American) influenced Korean youths acting out new life-styles in Seoul—that has occupied so much attention recently in Korea will be but one manifestation of a growing production of Korean culture that transcends its own national boundaries and comes back to shock its new middle-class sensibilities.

When the in-migration from other countries to Japan, Korea, Hong Kong, and Singapore is combined with the networks of transnational migrants from each of them, it is not too difficult to see that the future of cities in East Asia will be transformed. The assumed cultural homogeneity of some of these societies, particularly Japan and Korea, will have to be re-examined, as will the sources of influence of the diaspora from their own societies on the home social and political values and institutions in each. Although King (1993: 89) asserts that the increasing manifestation of multiculturalism and poly-ethnicity and the way they get represented in the built environment is subject to the 'rules, codes and policies of individual places and states', such statements beg the question of from where the 'rules, codes and policies' emanate. If current trends are any indication, the supposition that they emerge solely from within a given society cannot be so easily accepted. In any case, addressing multiculturalism and cultural diversity will be high on the social agenda in the twenty-first century.

The Commodification of Culture as the 'New' Comparative Advantage

Culture, whether it is through cultural icons embedded in the façades of commercial buildings and housing estates or in the form of cultural events created for international tourism and conventions, has received new attention as a leading factor in a city's economic success (Basset 1993). This form of culture, namely, its commercialization and commodification, has become the focus of post-modern critiques of the contemporary urban condition. Of course, the idea that cities would fabricate images to sell themselves is not new.[1] But in more recent times, the idea of filling urban spaces with 'instant' histories from any period of time, any place, or any real or imagined cultural roots has gone beyond selling real estate. It is now fundamental to a locality's comparative advantage in attracting investment in industry and business for economic growth. Inventing and playing with cultural 'artefacts' of the ages in the built environment has also more clearly moved into the realm of symbolic capital as part of the ensemble of luxury goods attesting to the command of history held by occupants and owners of commercial buildings and houses alike.

The emergence of commodified culture as comparative advantage is intricately linked to the global reordering of the geography of industrial capital over the past three decades. In the 1960s few urban and regional economists would have predicted the massive deindustrialization of industrial heartland regions that began in the USA and Europe in the 1970s. Nor would they have well understood the rise of new production spaces in hitherto unindustrialized areas that has accompanied the advent of a post-Fordist era of industrial and socio-political reorganization (Scott and Storper 1992).

As industry became increasingly internationally footloose, a region's comparative advantage in attracting industries shifted away from natural resource endowments and the concentration of large pools of industrial labour. Since almost any nation or locality is capable of building at least modest industrial estates and offering tax incentives to attract industry, the unique attractions in the selling of regions became cultural amenities and socio-economic environments that offered an attractive ambience for daily living, including perceived low levels of civil unrest. Regions could no longer promote themselves as smokestack sites as they did a half-century before. In place of industrial landscapes, they began to offer clean environments, diverse life-style opportunities—parks, museums, symphony

---

[1] In the words of Mayo (1933: 319; quoted in Davis 1992: 17) about one of the best-known examples: 'Los Angeles, it should be understood, is not a mere city. On the contrary, it is, and has been since 1888, a commodity; something to be advertised and sold to the people of the United States like automobiles, cigarettes, and mouth wash.'

orchestras, opera houses, historical sites, and frequent cultural events—as their comparative advantage.[2]

Rich cultural settings capable of acting as magnets for economic growth are not ubiquitous, and even when superficially fabricated, they may not be sufficient to attract and keep footloose enterprises. For underlying the physical representations of culture in urban space are more fundamental socio-cultural characteristics: the capacity to generate co-operative energies to engage in projects that enlarge shared amenity spaces, to maintain these spaces in their daily use, and to sustain the ambience of conviviality in them. As many Western cities seem to be finding out, the social and cultural dislocations visiting metropolitan regions in the late twentieth century may well be the sources of industrial decline in the twenty-first century. Deteriorating inner cities, urban crime, and violence are no longer simply social issues; they are becoming part of the economics of industrial flight and local economic decline. In East Asia, the increasing commuter times, traffic congestion, high levels of air pollution, and lack of green public space are already becoming important industrial location issues.

Commodified culture is playing an even more visible role in attempts to capture the growing international business service sector. Many nations and cities are now turning toward world-scale cultural events as a means of generating new sources of economic growth. Hosting the Olympics, World Cup, Asian Games, or becoming a site for international conventions or music and cultural festivals, is no longer a minor aspect of urban economies; for many they have become the mainstay. As the city of Seoul discovered in hosting the Olympic Games just a few years ago, the competition for the right to host this event and the capacity to carry it out were intensely cultural both in terms of the theatre of events put together for it and in maintaining the city-wide social atmosphere needed to guarantee its success. The recent failed attempt by Beijing to host the Olympic Games is another example. For many large cities in East Asia and the world in general, a perceived decline in tourism, which is packaged as an escape to exotic cultures, is seen as a major blow to their economies.

Such new dimensions of industrial location and economic growth may seem to be only relevant for the post-industrial economies of the USA and Europe, but there are many indications that they are becoming equally important in East Asia as well. With the possible exception of China and Vietnam, the economies of this region have reached a crossroads at which they must find alternatives to the labour-intensive factory system and the small industrial sweatshop if they are to sustain their economic growth in

[2] In the United Kingdom, for example, smaller towns were found to be attractive if they had such attributes as a restaurant with an entry in the *Good Food Guide*, a hotel in the *Michelin Guide*, antique shops, and access to a National Park (Glasson 1992).

the face of the rapid emergence of new sites for these activities in other parts of Asia, notably China.

Cities in Japan are perhaps most keenly aware of the compelling need to shift away from these older sources of economic growth. Competition among them to be selected as sites for major cultural events has reached new heights over the past two decades. Now, in any given year the list of expositions and festivals—international design expos, food festivals, flower expositions, just to name a few—is part of a concerted national government campaign to remake Japan into a 'Superpower of Life' aimed at displacing its image as 'Japan Inc.' In a more practical sense, cities in Japan know that a successful cultural event brings with it new infrastructure and amenities that can potentially carry them forward into a new phase of economic growth.

One of the more intriguing features of the shift toward social and cultural events is the rate at which the hosting of conventions has became an important component of urban economies in East Asia. Around the world one of the most prominent features of the 1980s was the substantial increase in investment in showcase convention centres—including hotel construction—and the hosting of international conventions. Interestingly, in terms of the number of conventions held, the city of Seoul moved from a world rank of 19 in 1982 to 11 in 1986, ahead of Tokyo, Hong Kong, and Taipei and only slightly behind Singapore.[3] Such events not only constitute a new direction toward local economic growth; they are part of an increasing global network of information exchange and learning that, despite the array of technologies they depend on, still thrive on face-to-face contact among people. The fact that people are now willing to travel thousands of miles for these encounters in 'exotic' locations only heightens the importance of culture in this new dimension of urban economic growth and development.

The cultural dimension of economic growth occupies what King (1993: 101) identifies as the 'second sphere' of the city as a site for accumulation of cultural capital that involves:

The massively expanding realm of the global cultural economy, the ideas, images, and signs which form an ever-increasing proportion of post-industrial economies and cultures: the information, advertising, and communication industries, the world of the fine and applied arts, film, video, TV, disc, tape, cable publishing—a sector which, in the economies of world cities, is increasingly oriented to foreign markets.

In other words, culture will be a principal export growth activity in the twenty-first century. East Asian cities have already become key sites for

---

[3] The top two cities held a reported combined total of 4,353 international conventions in 1982; this increased to 6,681 by 1986. Seoul held 36 in 1982 and 84 in 1986, moving it from 19th to 11th in rank. Singapore held 100 international conventions in 1986; Paris, the top-ranked city, held 358 (King 1993).

the production and export of culture. Hong Kong is one of the leading film producers in the world and has also become the site for television production and dissemination throughout Asia. The impact has been pervasive; even village children in remote rural areas of Asia can be found playing out *kung fu* dramas watched on television sets powered by automobile batteries.

There should be little doubt that centres where the dissemination of information is controlled will also be centres where culture-laden selections of information, which go beyond fine arts to include political and ideo-logical imagery, are made for others to consume. This realization is perhaps one reason why Singapore has announced its intention to rival Hong Kong as a world information centre. Such a move by Singapore brings into sharp focus the contradictions between, on the one hand, wanting to become a world city in an informational era while, on the other, clamping down on the flow of information and ideas within the city for political regime maintenance objectives.

Although economies may thrive on packaging culture for economic gain, the process falsifies as much as it authenticates cultural symbols and practices. Throughout the world, the packaging of culture for tour-ism is often a national self-parody of histories that never existed and cultural acts that were never practised. Moreover, the scaling up of cultural events and tours tends to replicate the same processes of transna-tionalization of control and ownership of major assets that have been criticized for generating few local multiplier effects other than low-wage hotel and theme park jobs. If cultural activities are to have meaning for the host society as daily practices, alternative models to the dominant forms of mass packaging of culture for commercial value will need to be promoted.

Culture as Resilience and Resistant

While many may portray culture as embracing societies in harmonic rela-tions, other perspectives point toward culture as a source of heightening discord and violence. There are at least two major dimensions of this view of culture. One concerns international relations and the reported emer-gence of very large regional cultural spheres—civilizations—that are said to be replacing political and economic differences as the major source of conflict among nations and societies. The second is the view 'from below', of culture being the organizing force against marginalization and opp-ression of the poor, minorities, and other victims of dominant cultures and localized structures of capitalist development.

With regard to East Asia, perhaps the most provocative exposition on culture as a fundamental cause of international friction is that presented by

Huntington (1993: 22), who in the venerable journal *Foreign Affairs* boldly declares that in the future: 'The great divisions among humankind and the dominating source of conflict will be cultural. Nation states will remain the most powerful actors in world affairs, but the principal conflicts of global politics will occur between nations and groups of different civilisations. The fault lines between civilisations will be the battle lines of the future.'

Defining civilization as a cultural entity, in that 'villages, regions, ethnic groups, nationalities, religious groups all have distinct cultures at different levels of cultural heterogeneity', Huntington argues that while politics and economics are negotiable, culture is less mutable and less easily compromised. As communications and transportation continue to shrink the globe, the butting of cultures against each other will intensify the conscious awareness of cultural differences. Moreover, the weakening of the nation-state as a source of identity and the spread of international cultural linkages through migration and other means has resulted in cultural identities transcending national boundaries to form region-wide affinities. This will in turn accentuate the much reported trend toward economic regionalism— fortress Europe, the Americas, and Pacific Asia.[4] The West is put forth in this milieu of increasing cultural consciousness as the source of confrontation and provocation against other civilizations, including a China-centred Confucianist one.[5]

Because it resembles a view being promoted by some Asian political leaders that Asian 'communitarian values' stressing social order over individual liberties are in the ascendance, Huntington's thesis has been received in East Asia with much acclaim as well as criticism (Ajami 1993; Mahbubani 1993; Harries 1994). As presented by East Asian leaders, this Orientalism-in-reverse, which presents a monolithic, uncontextuated view of East Asian societies, also serves to counter such recent modernization theses as that by Fukuyama (1992), who presents a linear view of history toward liberal democracy pioneered by the West and inevitably to be followed by the rest of the world.[6]

---

[4] In contrast to the current 'Look East' policy being promoted from Japan and with support from such countries as Malaysia that centre the emerging cultural-economic Asian region on Japan, Huntington and others (Harries 1994) proclaim that China, rather than Japan, will become the new epicentre of an East Asian civilization. Among the reasons for this assertion is the existence of extensive overseas Chinese networks in East and South-East Asia.

[5] In Huntington's words, 'the fault lines between civilisations are replacing the political and ideological boundaries of the Cold War as the flash points for crisis and bloodshed'.

[6] Fukuyama's unhistorical renderings of the Western experience and its current conditions—as well as flawed interpretations of Plato, Hegel, Nietzsche, and other theorists—have been widely noted (Gray 1992; Roth 1993). His thesis is fundamentally ideological in supporting the American neo-conservative view that capitalism and democracy not only go hand in hand, but, along with technology created by the innovative capitalist entrepreneur, have also solved all the major social issues of the Western world. As such it responds on the same plane as the 'Look East' ideology emerging in Asia.

There are, however, several key problems with this thesis. The most fundamental problem lies with its conflictual view, which denies the coexistence of Western and non-Western civilizations and, furthermore, the possibility of fusion between them. As pointed out by Mahbubani (1995), this fusion of Western and East Asian cultures underlies the explosive growth and democratization of East Asia in recent decades. Another major problem is the assumption that conflict between the stylized Confucian civilization and the equally stylized Western civilization will be consistently stronger than cultural differences and conflict within each (Harries 1994). This assumption cannot explain, for example, the profound distrust in Asia of Japan stemming from Second World War experiences; nor can it easily explain the many continuing separatist movements in Asia as well as in Europe and North America. Given, too, that neither Korea nor Japan is part of overseas Chinese networks, their position in the East Asia civilization in Huntington's mind is problematic at best. These observations suggest that culture and cultural struggles are much more localized than discourses at the level of 'civilization' allow.

The idea of culture as resistance at the global scale is further confounded by the second source of cultural divides, namely, what some have called the culture of the tea and coffee shop, or a popular culture emerging within and held by the marginalized and oppressed workers, minorities, and disenfranchised subgroups that are found within each society (Kuper 1994). A recently re-emerging theme concerning these segments of society is that this culture—born of perceived oppression and as a form of collective identity—is a central source of self-empowerment of the weak against the strong (Friedmann 1989, 1992; Evers 1985). One of the unanticipated features of the past decade in this regard has been the resilience and re-emergence of separatist movements based on cultural affinities throughout the world. None has been more unexpected and, in some cases, disheartening to the outsider than the rise of ethnic nationalism in the former Soviet Union and Eastern Europe. Equally dramatic has been the strengthening of First Nation—or so-called indigenous—peoples' organizations in Mexico, the USA, and many other quarters of the world. East Asia has its own such 'minorities': burakumin (an outcast group in Japan), Ainu, Okinawans, Koreans, and Chinese in Japan; Tibet and other regions in China; and Malay and Indian populations in Singapore, to name a few of the better-known examples.

Most of these popular movements have been targeted not at Western civilization, but at the more immediately perceived source of oppression—the state (Apter and Sawa 1984). In Latin America in the 1960s and 1970s, for example, the leftist movements brought to their support artists, writers, musicians, and other craftsmen of cultural symbols to assist in presenting a liberating, egalitarian vision of Latin America's future.[7] As explained by NACLA (1994: 15), the focus on culture was obvious:[8]

Culture is the symbolic realm where people derive their values, beliefs and ideals at a certain place and time. It is difficult to separate politics and culture since politics always takes place in a cultural environment. It is within the realm we call culture that we get our bearings in life; it is there that we ingest the notions of what is good, bad, just, natural, desirable and possible. Politics takes place in a cultural milieu; to inquire into the political uses of culture is to inquire into the politics of hope and desire—and, of course, fear.

East Asia has seen its own versions of social movements against both dominant and 'high' culture. Kwangju in South Korea continues to be a significant source of anti-statist social mobilization. As noted, each of the other settings explored in this book—Hanoi, Singapore, Hong Kong—is itself multiracial and multi-ethnic, making assertions of ethnic or racial homogeneity dubious.

Complicating all of the resurgence of cultural identity as the source of resistance is international migration. The United Nations reported that in 1993 there were at least 100 million people who lived outside their country of birth or citizenship (Williams 1994). With about one-fifth being classified as refugees, the remaining 80 million were mostly legal and illegal migrant workers. As previously noted, the portent of multicultural societies in East Asia brings with it the potential for politics of resistance to either or both incorporation into and continued suppression and discrimination by the dominant cultural/ethnic group. The struggles of Koreans in the 1980s in Japan to stop being fingerprinted as a condition for being given permanent residence cards are but one of the least violent examples of these politics.

Rather than a clash of civilizations, the world is becoming much more complex, with overlays of international movements of people, capital, and information clouding neat world divisions. The world is indeed becoming more interdependent through the combination of instant information (the power of the fax machine has been amply demonstrated in the 1989 Tiananmen incident), economic interdependence, and the appeal of free

---

[7] NACLA concludes (1994: 15), however, that: 'Now, in the current era of Left retrenchment, the printed word, and the literary culture which grew up around it, is being encroached upon by the commercial music industry and television. As the powerful workings of the "free market" are presented as natural and inevitable, economists are replacing the literary intelligentsia as guardians of a very different kind of utopian flame. Culture, its products and its media of dissemination are increasingly commodified. The wealthy and the powerful have captured the citadels of culture—but not everywhere. . . . The struggle continues, and a good part of it is being fought on cultural terrain.'

[8] Implied in this recognition of the role of culture in social mobilization is the view that these movements often mask the principal source of social cleavage, namely divisions on economic class, with cultural and ethnic or even regional identities. Calling upon cultural practices, beliefs, and values is easier than appeals to emergent class divisions in mobilizing oppressed groups. Similarly, violent reactions against the oppressed and the poor are also often couched in ethnic, racial, or cultural terms.

markets, which in turn suggests the possibility of hybrid cultures that are a blend of Western and non-Western cultures.

How do the above observations relate to the city? First, cultural clea-vages—whether based on ethnicity, race, regional origin, or class—within cities can be expected to become more salient in the future as urban populations in Asia become more multicultural. Second, the strands of multiculturalism will also be increasingly 'internationalized' and difficult to either influence or control from within nations and localities. Third, the particular intense agglomeration of activities and the form of the built environment of cities—historical landmarks, skyscrapers bringing thou-sands of people together in conspicuously high densities in buildings that are themselves symbols of power—will make the city the site for both symbolic appeal and symbolic violence. Cities such as Beijing, Seoul, and Tokyo are celebrated as symbols of a cultural continuity that has persisted through the ages even in the face of social upheavals, economic transformations, and technological revolutions. The juxtaposition of the temple with the finance centre, the palace with the parliament, and the ancient city gate with the modern expressway tollgate is portrayed as representing a cultural resilience that intertwines rather than displaces the ancient with the modern.

On the other hand, cities are subject to symbolic violence and manipula-tion in a massive proportion. Whether it is the massacre in Tiananmen Square in Beijing, the bombing of the World Trade Center in New York, the gas poisoning of commuters in the subways of Tokyo, or even the explosion of a federal building in Oklahoma, the Cold War era of preparing for wars between nations has moved into a post-Cold War era that is already marked by new forms of extreme violence for which there are few historical precedents. Whether or not this can be contained and what type of socio-cultural and political institutions will best work to mediate potentially volatile social and cultural cleavages will be among the most important questions that cities will have to address everywhere.

Culture as a Basis for Societal Guidance

Fundamental economic restructuring, world-wide movements for political reform that are confronting the very basis of state authority, and increasing cultural diversity in cities everywhere all indicate that the twenty-first cen-tury will be a new era in thinking about and taking action to improve and sustain urban societies. Efforts to secure the longer-term sustainability of cities in this new era will rest greatly on the capacity to generate new ways of governance and taking collective action in each national and local setting. Enlarging this capacity will in turn call for the reconstitution of cities themselves in an institutional as well as physical sense.

What would 'an appropriate space' (Lefebvre 1991: 59) for reconstituted urban society entail? Much depends upon the particular contexts, but in East Asia it would point toward: a built environment for public discourse and civic affairs accessible to citizens at large; an accommodation of diversity and multicultural landscapes; equitable access to housing, land, and work; expanded public spaces and amenities; environmental infrastructure and services for clean cities; and new spaces for culture-related economic base activities. Taken together, variations of this list present a formidable agenda for the future of cities. A central question with regard to cultural change and the city is whether society can (re-)establish a cultural basis in shared values that will allow visions of the future to be created and acted upon in a manner that addresses major issues and dilemmas. As a corollary, we can ask whether indigenous cultures and local traditions can provide sources for the establishment of such a cultural basis.

As observed in Singapore's renewed emphasis on the cultural dimension of urban life, each East Asian society is searching for a model for its future. Rich cultural traditions in East Asia such as Confucianism, Buddhism, and Taoism, if grafted on the universal values of Western culture, may provide alternative models of development as well as bring back cultural and ethical dimensions to the discussion of the urban future in East Asia. In these hybrid outcomes, we may find ways and means to transcend the dichotomy of communitarianism versus individualism and to strike a balance between morality and materialism.

More directly for the cities, the question to be asked is to what extent can new forms of societal guidance and social mobilization (Friedmann 1987) be created through new or transformed socio-cultural and political institutions? Much of recent literature on the city in the USA and Europe has taken a decidedly pessimistic position about the prospects for social co-operation to emerge in the coming years.[9] To many writers the future is already foretold, and it is not a promising one. In this vein, Davis (1992: 3) provocatively states that 'the best place to view Los Angeles of the next millennium is from the ruins of its alternative future'. In other words, the path that could have been taken has been so diverted by events of the past few decades that to study the city now is to engage in the archaeology of opportunities misspent and futures gone awry. Similar observations have led others to conclude that in a post-modern world in which grand social projects and visions are dismissed as modernist pretensions, planning has 'lost its way' (Knox 1993: 11):

Not so long ago, the planning profession was carried forward by visionary ideals and a conviction that goodwill and technical competence could achieve lasting

---

[9] The spectre of socio-cultural disintegration, economic decline, and catastrophic environmental collapse haunts current interpretations of cities and their future throughout the world. A perusal of recent writings about the plight of cities in North America is not comforting (Harrison and Bluestone 1988; Smith and Feagin 1989; Davis 1992; Harvey 1992; Knox 1994).

improvements in urban life. It did not take long, however, to discover that planning could not deliver the utopian goods . . . Planning practice became estranged from planning theory and divorced from any broad sense of the public interest . . . It became increasingly geared to the needs of producers and the wants of consumers and less concerned with the overarching notion of rationality or criteria of public good.

For others writing from a similar perspective, the decline of the social capacity for planning in the public realm has been a product of late capitalist development that has compelled local governments to opt for expedient short-term economic growth initiatives over redistributive programmes, education, and other strategic long-term development choices. This rise of the entrepreneurial local state and capitulation to profit have undermined the capacity to envision the city as what Friedmann (1988) calls a 'life space' as well as a vortex in the global flow of capital. As concluded by Beauregard, 'The result has been a peculiar form of nonplanning in which planners participate in individual projects, often attempting to temper the most egregious negative externalities, while failing to place these projects in a broader framework of urban development . . . Planning has become entrepreneurial and planners have become dealmakers' (1989: 387–8).[10]

Finally, the critique of contemporary urban planning points toward systemic global forces that are generating many of the pressures impinging on cities (Glickman 1989). The phenomenal collapse of space–time relations made possible by revolutionary advances in transportation and communications and the emergence of corporate entities capable of taking advantage of them on a global scale has made these pressures all the more invasive as well as pervasive. From this perspective, cities are compelled to 'bid down' in their competition for economic growth by offering ever higher benefits to would-be investors in return for diminishing rewards. In the process, the capacity to translate economic growth into a socially and environmentally balanced urban habitat is severely diminished (Harrison and Bluestone 1988). These 'pressures from the global economy' are identified as a principal source of structural unemployment, worsening working conditions at the bottom of the employment ladder, and increasing urban poverty (Goldsmith and Blakely 1992).

While each of these views has merit they are none the less partial and are subject to caution when applied to East Asia. They are partial even in the US context. Enlightened and progressive city governments do exist. Not all cities are suffering the same problems that are ascribed to the general case. Moreover, the state and the private sector are not the only actors in urban planning, management, and development. Grassroots movements, nongovernment organizations, and a host of other forms of association outside

---

[10] Quoted in Knox (1993: 12).

business and government have greatly expanded to become major sources of assistance and economic support, including the promotion of micro-enterprises in low-income and minority neighbourhoods.

More to the point of the discussion here, global forces and the malaise of US cities do not either separately or together constitute a future portrait of cities in Asia. Among the key intervening variables that must be reinserted into the understanding of societal interventions into processes of urbanization is culture—not merely in the sense of the arts and popular festivals, but in the sense of values, codes, and practices that become represented in social and political relations and, ultimately, in the building of cities and the animation of daily life in them. Without this understanding, urban studies become ahistorical exercises of imposing generalized experiences onto every locality.

In the context of global forces and cultural change, some theorists have argued that the increasing penetration of international capitalism into national and local economies will lead to a 'denationalization' of history (Agnew and Duncan 1989; Ohmae 1995) and a 'rendering of the world as a single place' (King 1989: 5; Wallerstein 1991). But a fuller understanding of development that allows for real histories to be played out within each society sides with alternative views put forth by King (1993) and others (Robertson 1987; Featherstone 1991), which argue that despite the persistence of international forces impinging on local culture, there is little prospect for a single global culture. There will instead be many types of cultural transformations emanating from global–local interaction.

Nowhere has the localization of global 'imperatives' (Storper and Walker 1989) been more striking or more important in the thinking about development than in East Asia.[11] Over the course of the past four decades, a small number of East Asian societies has surprised the world by moving from being among the poorest to joining the most economically advanced in the world through strategies directly linking their fortune with the international economic system. Over the last decade they have surprised the world again by carrying out political reforms thought improbable by many only a few years earlier.

These economic advances and political reforms are not simply the outcome of global economic forces choosing to develop a few exemplars; nor are they the product only of a 'developmental state'. They most fundamentally involve cultural and social institutions that conditioned ways in

---

[11] These imperatives emanate from the requirement to sustain processes of accumulation through international competition. In the past, this requirement was met in East Asia through a very high level of state intervention, first, to maintain steady supplies of low-wage labour while rapidly increasing the productivity of labour; second, to invest heavily in economic and social overhead capital (such as housing) needed to maintain high levels of international competitiveness and socio-political stability; and, third, with the exception of Singapore, to assist in building the basis for the growth of domestic firms by directing finance and other resources to them (Douglass 1994).

which people collectively complied with as well as struggled against authoritarian political structures. Multiculturalism (pluralism), commodification
of culture, cultural renewal and resistance, and reformed state–civil society
relations that will occupy much of the urban agenda in the coming years
underscore the many ways in which transformations in city life are not
wholly local, national, or international in origin, but are instead outcomes of
interactions across all these scales. They are at the same time part of the
dynamics that intertwine cultural institutions with the structuring of power
and the physical production of urban space.

References

Agnew, John A., and Duncan, James (1989). Introduction, in John Agnew and
James Duncan (eds.), *The Power of Place*. Boston: Unwin Hyman.
Ajami, Fouad (1993). 'The Summoning', *Foreign Affairs*, 72: 2–9.
Apter, David, and Sawa, Nagayo (1984). *Against the State*. Cambridge, Mass.: Har-
vard University Press.
Basset, K. (1993). 'Urban Cultural Strategies and Urban Regeneration: A Case
Study and Critiques', *Environment and Planning A*, 25: 1773–998.
Beauregard, Robert (1989). 'Between Modernism and Postmodernism: The Ambiv-
alent Position of U.S. Planning', *Society and Space*, 7: 381–96.
Davis, Mike (1992). *City of Quartz: Excavating the Future in Los Angeles*. Los
Angeles: Vintage.
Douglass, Mike (1992). 'Global Opportunities and Local Challenges for Regional
Economies', *Regional Development Dialogue*, Summer (13): 3–21.
——(1994). 'The "Developmental State" and the Newly Industrialized Economies
of Asia', *Environment and Planning A*, 26: 543–66.
Evers, Tilman (1985). 'Identity: The Hidden Side of New Social Movements in Latin
America', in D. Slater (ed.), *New Social Movements and the State in Latin Amer-
ica*. Amsterdam: Centre for Latin American Research and Documentation.
Fainstein, Norman (1993). 'Race, Class and Segregation: Discourses about African
Americans', *International Journal of Urban and Regional Research*, 17 (3): 384–
403.
Featherstone, Mike (1991). *Consumer Culture and Postmodernism*. London: Sage.
Friedmann, John (1987). *Planning in the Public Domain: From Knowledge to Ac-
tion*. Princeton: Princeton University Press.
——(1988). *Life Space and Economic Space*. New Brunswick, NJ: Transaction
Books.
——(1989). 'Collective Self-Empowerment and Social Change', *Ifda Dossier*, 69: 3–
14.
——(1992). *Empowerment: The Politics of Alternative Development*. New York:
Basil Blackwell.
Fukuyama, Francis (1992). *The End of History and the Last Man*. New York: The
Free Press.

Glasson, John (1992). 'The Fall and Rise of Regional Planning in Economically Advanced Nations', *Urban Studies*, 29 (3–4): 505–31.

Glickman, N. (1989). 'Cities and the International Division of Labor', in M. P. Smith and J. Feagin (eds.), *The Capitalist City*. New York: Basil Blackwell.

Goldsmith, William, and Blakely, Edward (1992). 'Separate Opportunities: The International Dimension of American Poverty', in *Separate Societies: Poverty and Inequality in U.S. Cities*. Philadelphia: Temple University Press.

Gray, John (1992). 'The End of History and the Last Man' (book review), *National Review*, 44 (9): 46–9.

Harries, Owen (1994). 'Power and Civilization', *National Interest*, Spring (35): 107–13.

Harrison, B., and Bluestone, B. (1988). *The Great U-Turn: Corporate Restructuring and the Polarization of America*. New York: Basic Books.

Harvey, David (1989). *The Condition of Postmodernity*. London: Basil Blackwell.

——(1992). 'Social Justice, Postmodernism and the City', *International Journal of Urban and Regional Research*, 16: 587–601.

Hays, Michael (1995). 'Representing Empire: Class, Culture and the Popular Theatre in the Nineteenth Century', *Theatre Journal*, 47 (1): 65–83.

Huntington, Samuel P. (1993). 'The Clash of Civilizations', *Foreign Affairs*, 72 (3): 22–50.

Javier, Gideon (1993). 'Filipino Day Laborers in Yokohama's Kotobuki-cho', Seminar on Foreign Workers in Japan: Gender, Civil Rights, and Community Response. East–West Center, 1–3 Dec. 1993, Honolulu.

Johnson, James, *et al.* (1993). 'Seeds of the Los Angeles Rebellion of 1992', *International Journal of Urban and Regional Research*, 17 (1): 115–19.

Jones, Delmos (1992). 'Which Migrants? Temporary or Permanent?', in Nina Schiller *et al.* (eds.), *Towards a Transnational Perspective on Migration: Race, Class, Ethnicity, and Nationalism Reconsidered*. New York: Annals of the New York Academy of Sciences.

Kim, Won Bae (1995). 'Regional Interdependence and Migration in Asia', *Asia and Pacific Migration Journal*, forthcoming.

King, Anthony D. (1989). 'Colonialism, Urbanism, and the Capitalist World Economy', *International Journal of Urban and Regional Research*, 13: 1–18.

——(1993). 'Identity and Difference: The Internationalization of Capital and the Globalization of Culture', in Paul Knox (ed.), *The Restless Urban Landscape*. Englewood Cliffs, NJ: Prentice Hall.

Knox, Paul (1993). 'Capital, Material Culture and Socio-spatial Differentiation', in Paul Knox (ed.), *The Restless Urban Landscape*. Englewood Cliffs, NJ: Prentice Hall.

——(1994). *Introduction to Urban Geography*. Englewood Cliffs, NJ: Prentice Hall.

Kuper, Adam (1994). 'Culture, Identity, and the Project of a Cosmopolitan Anthropology', *Man*, 29 (3): 537–55.

Lefebvre, Henri (1991). *The Production of Space*. Oxford: Basil Blackwell.

Machimura, Takashi (1993). 'How Foreign Workers are Incorporated into Local Communities: Three Types of Socio-spatial Clustering in Greater Tokyo', Seminar on Foreign Workers: Gender, Civil Rights, and Community Response. East–West Center, 1–3 Dec. 1993, Honolulu.

Mahbubani, Kishore (1993). 'The Dangers of Decadence', *Foreign Affairs*, 72: 10–14.

——(1995). 'The Pacific Way', *Foreign Affairs*, 74: 100–11.

Marcuse, Peter (1993). 'What's so New about Divided Cities?', *International Journal of Urban and Regional Research*, 17 (3): 355–65.

Mayo, Morrow (1993). *Los Angeles*. New York.

NACLA (1994). *NACLA Report on the Americas*, Sept.–Oct. 28 (2): 15.

Ohmae, Kenichi (1995). *The End of the Nation-State*. New York: The Free Press.

Pang, Eng Forng (1993). 'Regionalisation and Labour Flows in Pacific Asia', Development Center of the Organization for Economic Co-operation and Development, Paris.

Pred, Alan (1986). *Place, Practice, and Structure: Social and Spatial Transformation in Southern Sweden, 1879–1950*. Totowa, NJ: Barnes & Noble.

Robertson, R. (1987). 'Globalization Theory and Civilization Analysis', *Comparative Civilization Review*, 17: 20–30.

Roth, Michael (1993). 'The End of History and the Last Man' (book review), *History and Theory*, 32 (2): 189–97.

Schiller, Nina, *et al.* (1992). *Towards a Transnational Perspective on Migration: Race, Class, Ethnicity, and Nationalism Reconsidered*. New York: Annals of the New York Academy of Sciences.

Scott, A. J., and Storper, M. (1992). 'Regional Development Reconsidered', in E. Huib and V. Meier (eds.), *Regional Development and Contemporary Industrial Response*. London: Belhaven.

Smith, Michael P. (1992). 'Cities after Modernism', in M. P. Smith (ed.), *After Modernism: Global Restructuring and the Changing Boundaries of City-Life. Comparative Urban and Community Research*, 4.

——and Feagin, J. (eds.) (1989). *The Capitalist City*. New York: Basil Blackwell.

Soja, Edward (1989). *Postmodern Geographies*. London: Verso.

SOPEMI (1993). *Trends in International Migration; Annual Report 1993*. Paris: OECD.

Storper, Michael, and Walker, Richard (1989). *The Capitalist Imperative: Territory, Technology, and Industrial Growth*. New York: Basil Blackwell.

Wallerstein, I. (1991). 'The National and the Universal: Can There Be Such a Thing as a World Culture?', in Anthony King (ed.), *Culture, Globalization and the World System*. Binghamton, NY: Department of Art and Art History, SUNY Binghamton.

Williams, Robin M., Jr. (1994). 'The Sociology of Ethnic Conflicts: Comparative International Perspective', *Annual Review of Sociology*, 20: 49–81.

INDEX

Philippines (*cont.*):
  women in the workplace 79
  see also Manila
politics 97, 176–7, 192, 246
  aloofness 102, 195–6
  change 175–7
  national ideology 125–50, 155
  order, pre-modern East Asia 24–6
population 75
  Beijing 18, 73, 133, 135–6, 138
  Hanoi 173, 179
  Seoul 17, 25
  Tokyo 17, 32, 157
poverty 114, 115
power 47, 52, 159
  allocative and coercive 234–5
  appropriation and concentration 7
  compromised 58–9
  constitutional 189
  economic 107, 236
  function, symbolism and 7–9
  indigenous élites 51
  institutions which can shape regulation
    of 5
  political 236, 237
  structuring of 253
  symbolism 109, 247
privatization 10, 11, 17, 161, 177
prostitution 81, 113, 117, 238
Pusan 115 n., 120

Rangoon 50
real estate 160, 180–1, 192–5
resilience 243–7
resistance 96, 101, 243–7
restructuring:
  cities/urban 51, 52, 58, 152–3
  economic 10
  industrial 138
Rhee Seung-man 115, 117
Riau 18
Russia 110
  see also Soviet Union
Russo-Japanese War (1904–5) 156

Saarinen, Eliel 134
Saigon 29, 50
samurai 17, 24, 25, 32
San Francisco Peace Treaty (1952) 159
Seoul 8, 9, 21, 25, 50
  built environment 98, 99, 100, 109
  Confucian-influenced patriarchal values
    79
  golf courses 85–6
  high primacy 25
  housing 84, 85
  Japanese colonial rule 31, 109–15
  minority subcultures 6
  modernism and development 104–24

modified version of political centre 29
old city markets 81
orthogenetic tension 4
population 17, 25
pre-industrial, formation and development 105–9
Sae'oon Sang'ga and Yoido projects 99, 105, 115–23
urban structure and colonialism 28
world rank 242
see also Hanyang; Kim Hyun-Ok
separatist movements 245
Shanghai 9, 50, 82, 193
Shenzhen 197
Shinto 117
Singapore 34, 176, 177, 191
  administrative and managerial jobs 80
  built environment 98, 99, 100, 101, 212–33
  colonialism 6, 23, 29, 33, 50, 213–16
  communitarian political culture 10
  Confucian-influenced patriarchal values 79
  economic restructuring 10
  export-oriented industrialization 59
  heterogenetic tension 4
  housing 84, 85
  implicit developmental models 55
  industrialization 220
  intention to rival Hong Kong as a world
    information centre 243
  low-cost labour 54
  Malay and Indian populations 245
  Malays marginalized in 216
  migration to 81, 239
  night markets 81
  ownership of selected consumer durables 83
  population growth 75
  productive employment opportunities 70
  remarkable economic growth 8
  social engineering 18
  'transborder' urban regions 58
  urban transition 73, 74
  women's labour force participation 78
  world rank 242
  youth labour participation 76
  see also Lee Kuan Yew
Singapore river 225–6
Sino-Japanese War (1937–45) 158
social change 7, 11, 45–8, 50–5, 60, 177–8
  significant 235
social engineering 18, 34
social order 6, 31, 108
  hierarchical 27, 34
  nurtured by *samurai* culture 17
  pre-modern East Asia 24–6
  stratified 21
  suffocating 29

Index compiled by Frank Pert